MICHAEL HARR · RAJIV KOHLI

Commercial Utilization of Space

An International Comparison of

Framework Conditions

BATTELLE PRESS

Columbus · Richland

Library of Congress Cataloging in Publication Data:

Harr, Michael, 1947-
    Commercial utilization of space: an international comparison of frame-
work conditions/Michael Harr, Rajiv Kohli.
    Bibliography.
    ISBN 0-935470-46-8. $44.50.
    1. Space industrialization. 2. Reduced gravity environments—Industrial
applications—Economic aspects. 3. Remote sensing—Industrial applications—
Economic aspects. I. Kohli, Rajiv, 1947-. II. Title.
    HD9711.75.A2H37   1989
    338.0919—dc19
    Library of Congress Number: 88-7471

Additional copies may be ordered through Battelle Press, 505 King Avenue,
Columbus, Ohio 43201, 614/424-6393.

Preface

Since the early 1980s, considerable efforts have been made to develop space infrastructure in the United States, Western Europe, Japan, the Soviet Union, and in an increasing number of other countries. Although the development of space activities was initially motivated by reasons of national prestige, military leadership, and space science research, the utilization of space for commercial purposes is now assuming greater importance. Of the major space utilization segments of telecommunications, navigation, and earth observation, as well as the utilization of weightlessness, the telecommunications segment has already been commercialized. By contrast, earth observation is being operated commercially to a small extent, while the utilization of the microgravity of space has yet to attract any significant industrial activity. This is unfortunate because it is the latter segment that is said to hold considerable commercial potential in the future.

This commercial potential together with the efforts to share the very high entry costs for space utilization with the private sector has led all Government space agencies in the Western World to actively solicit the participation of the private sector in the utilization of space, and thereby achieve the major goal of transforming space programs from the status of national development programs to self-supporting financial ventures. It is in the interest of the individual countries to attempt to ensure that the domestic industry maintains a technological lead over foreign competition, but at the very least, to establish conditions which are favorable to domestic industry. Hence, the planning for the development of space infrastructure with Government funding must begin with an analysis of the existing framework conditions for private industry within the individual countries.

As a result, in 1986 the Ministry of Economics of the Federal Republic of Germany contracted with Battelle Institut in Frankfurt to compare and analyze the framework conditions in the major space nations in the Western World: United States of America; Western Europe with Germany, France, Italy, and the United Kingdom; and Japan.

An important requirement for this project was that the research team possess extensive knowledge of the conditions in the countries to be studied. Therefore, Battelle Frankfurt assembled a project team which included experts from Battelle Columbus for the United States and Mitsubishi Research Institute for Japan.

An international comparison of national framework conditions for the commercial utilization of space can start with only limited parallelism of the empirical criteria in the individual countries. In fact, even the objectives or the substance of similar institutional, legal, political, and other regulations in the individual countries for political or historical reasons have often been developed in different forms. Sometimes similar regulations even have different interpretations in the individual countries. Thus, the emphais in this work is primarily on illustrating the interactions of the framework conditions for potential customers of utilization of space.

This study first characterizes the nature and form of industrial space utilization projects and the progressive sequence of the individual steps of the project. From this information, the demands of private industry on the framework conditions are derived. These demands are then compared with the existing framework conditions in the individual countries to see how well these are met.

The present book is based on the German report *Internationaler Vergleich der Rahmenbedingungen für die kommerzielle Nutzung der Raumfahrt insbesondere in den Bereichen Mikrogravitation und Erderkundung,* presented to the Ministry of Economics in June 1987. It was approved for publication by the Minister of Economics on December 14, 1987. The information presented in this book is current through the Spring of 1989. Hence, nonspecific time phrases throughout the text such as "until now" and "to date" are referenced to that time.

Acknowledgements

Special thanks are extended to the project managers in the Ministry of Economics, Dr. Werner Friske and Dr. Franz-Josef Mathy, who not only initiated this work, but also provided critical reviews and comments during many long discussions with the project team. This book could not have been written without the contributions of the research team to the German report. The authors are especially grateful to their colleagues Dr. Irmtraud Kramer, George Mourad, Ingrid Schubert, Dr. Wolfgang Steinborn, Dr. Josef Trischler, Dr. Wolfgang Ulrici, and Dr. M. Ishikawa for their contributions, and to Karen Rush and Renate Hintze for assistance with manuscript preparation. Lastly, the book would never have been published without the encouragement of Joseph Sheldrick of Battelle Press and the able assistance of our editor, Cherlyn Paul.

Table of Contents

LIST OF TABLES

LIST OF FIGURES

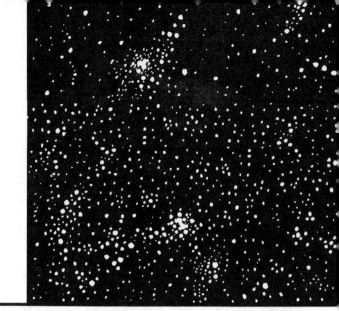

Commercial Utilization of Space

An International Comparison of

Framework Conditions

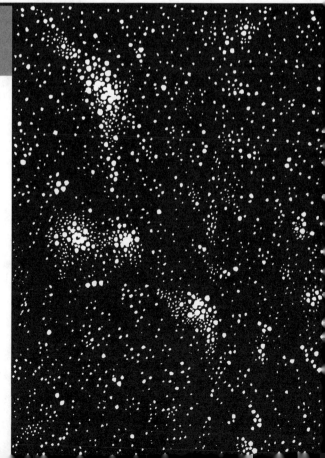

CHAPTER

1 Introduction

Commercial activities in space are emerging as a potentially important segment of the economy in the Western World. Taken as a whole, the space sector of the economy consists of four interrelated segments: transportation, satellite communications, satellite remote sensing, and research, development, and eventual production of materials under microgravity conditions in low Earth orbit (LEO). The involvement of the private sector in commercial space utilization is of interest to national governments for several reasons. First, advanced technologies already developed and those yet to be developed in space sectors will lead to innovations in the major economic sectors and, in turn, to the creation of new jobs. Second, industry as a whole can increase its competitiveness in world markets by developing and applying space-based advanced technologies.

These basic issues are evident in commercial space activities in the Federal Republic of Germany, United States, France, Italy, United Kingdom, and Japan. However, the emphasis in microgravity utilization and remote sensing activities varies widely among these countries.

Space activities in these four segments are not conducted under free market conditions in any industrialized nation. The sizable financial investment required and the uncertain profit potential make the development of space systems simply too risky for private industry. The risk will not decrease in the foreseeable future.

All of the countries considered support commercial space ventures through development of and access to the essential technical infrastructures, as well as by direct financial support such as providing low-cost flight opportunities. Hence, a variety of different framework conditions have been created for private sector utilization of commercial space opportunities that can improve or restrict the competitive position of national industries. To develop future policy options, national and international commercial space infrastructures must be understood. The institutional, legal, and political framework conditions, as well as restrictions and supporting initiatives for commercial space utilization in the individual countries are discussed and compared at length.

International comparison of the existing conditions for commercial utilization of space is based on similarities in the empirical regulations in the different countries. Often, for historical and political reasons, similar regulations with seemingly identical objectives are interpreted differently in different countries. As a result, attention is focused on the effectiveness of the existing conditions from the viewpoint of a private company interested in space utilization.

For a space venture to be commercially attractive to the potential user, it must meet certain requirements. The commercialization initiatives developed by national governments are designed to create the conditions necessary to meet such requirements. Thus, the requirements for typical commercial space utilization are first defined, then commercialization initiatives of individual countries are discussed and compared.

1

The technical development of satellite remote sensing and of microgravity utilization is very different, as is the nature of the commercial projects in these segments. These dissimilarities result in very different requirements and, correspondingly, in different conditions necessary to meet those requirements. We found it useful to treat these topics separately, a division clearly reflected in the discussion of satellite remote sensing (Part 1) and microgravity (Part 2).

For commercial satellite remote sensing, several barriers have led to different requirements at different stages of a utilization scenario for a commercial venture:

- Operation of remote sensing satellites
- Acquisition and dissemination of remote sensing data
- Commercial use of the data.

For microgravity utilization, these scenario stages are:

- Operation of the carrier systems
- Fabrication and distribution of experimental hardware
- Utilization of systems and hardware.

The major barriers to successful commercial microgravity utilization presently exist at the third stage: the absence of specific proposals for commercial utilization of existing carrier systems and hardware. Hence, this stage is emphasized in comparing the existing conditions in microgravity utilization, particularly as related to the existing institutional infrastructure available to private industry.

The countries considered differ significantly in political and economic conditions, in institutional infrastructure, and in the pace and nature of the development of space activities. As a result, detailed information on space activities was obtained for the existing conditions within Western Europe, the United States, and Japan. Published information was supplemented by interviews with key government and industry personnel. This information was compared and analyzed to determine the effectiveness of the existing conditions to the potential commercial user.

PART

1

Conditions for Commercial Utilization of Space for Satellite Remote Sensing

CHAPTER

2 Significance of Satellite Remote Sensing

When assessing the suitability of remote sensing from space, it must be compared with long existing conventional remote sensing techniques using aircraft and balloons, as well as with terrestrial methods for obtaining corresponding data. Most remote sensing data is still obtained using the latter methods, although the advantages of space-based sensing are becoming increasingly evident [1,2]:

- Data can be obtained simultaneously over a wide range of selected spatial scales.

- Remote sensing is independent of political divisions.

- Sensing is possible of remote and difficult terrain, which may be otherwise inaccessible.

- The data can be continuously actualized with relatively little effort.

- Satellite remote sensing provides repetitive coverage over extended time periods, and the reproducibility of imaging conditions for photographs is excellent.

- The use of radar imaging systems allows data acquisition regardless of weather conditions or time of day.

On the other hand, digital storage of remote sensing images and their subsequent routine computer manipulation are not inherent advantages of satellite imaging, as has been suggested [1,3]. The disadvantages of satellite remote sensing over aerial photography are

- The spatial resolution is smaller as Table 1 shows. For example, the data from the Landsat-MSS is suitable only for maps with a coarse scale of 1:200,000. For topographic maps, land use applications, and detailed forest destruction mapping, the resolution must be significantly higher. For military reasons, higher resolution data will not be made available from the United States or France [4].

- Satellites follow a fixed orbit and flight plan, providing essentially instant images. Dynamic processes are difficult to follow continuously because the flight plan cannot be modified to accommodate such events.

- The viewing angle is often highly oblique.

The principal applications of remotely sensed data are:

- Weather and climate observations

POTENTIAL FOR UTILIZATION

5

- Marine transportation and ship routing

- Offshore oil and gas exploration and extraction

- Deep sea mining

- Marine fishing

- Arctic observations

- Sea ice observation and monitoring

- Geological and geophysical information

- Environment and pollution monitoring

- Agriculture

- Land observation

- Location of inland water resources for irrigation planning

- Cartography

- Wildlife monitoring and management

- Forestry

- Crustal movements and earthquake research

- Monitoring renewable and nonrenewable resources

- City planning and development

- Coastal zone management.

TABLE 1. *Data on Civilian Commercial Remote Sensing Satellites*

Satellite	Repeat cycle	Spectral bands (μm)	Resolution (m)
Landsat-4 and Landsat-5	16 days	0.45- 0.52	30
		0.52- 0.60	30
		0.63- 0.69	30
		0.76- 0.90	30
		1.55- 1.75	30
		2.08- 2.35	30
		10.40-12.50	120
SPOT-1	26 days (2 to 5 days with pointing of sensors)	0.50- 0.59	20
		0.61- 0.68	20
		0.79- 0.89	20
		0.51- 0.73	10

The potential importance of remote sensing in these applications is illustrated by the examples below. Battelle has recently analyzed the demand in West Germany for remote sensing in these applications to assist in government planning [5].

- **Agriculture.** Detailed large-scale photographs can be used to obtain complete and timely information about crop acreage and yields, crop harvesting, and planting based on soil conditions and water availability. Large-scale crop damage by chemical pollutants or organisms can be easily detected [6]. Figure 1 shows the application of remote sensing data in the promotion phase of a typical agricultural development project.

- **Forestry.** Entire forest areas can be mapped and observed to prevent destruction and to gather information on forest statistics. Classification of forest destruction trends, however, requires a resolution of 10 meters or less.

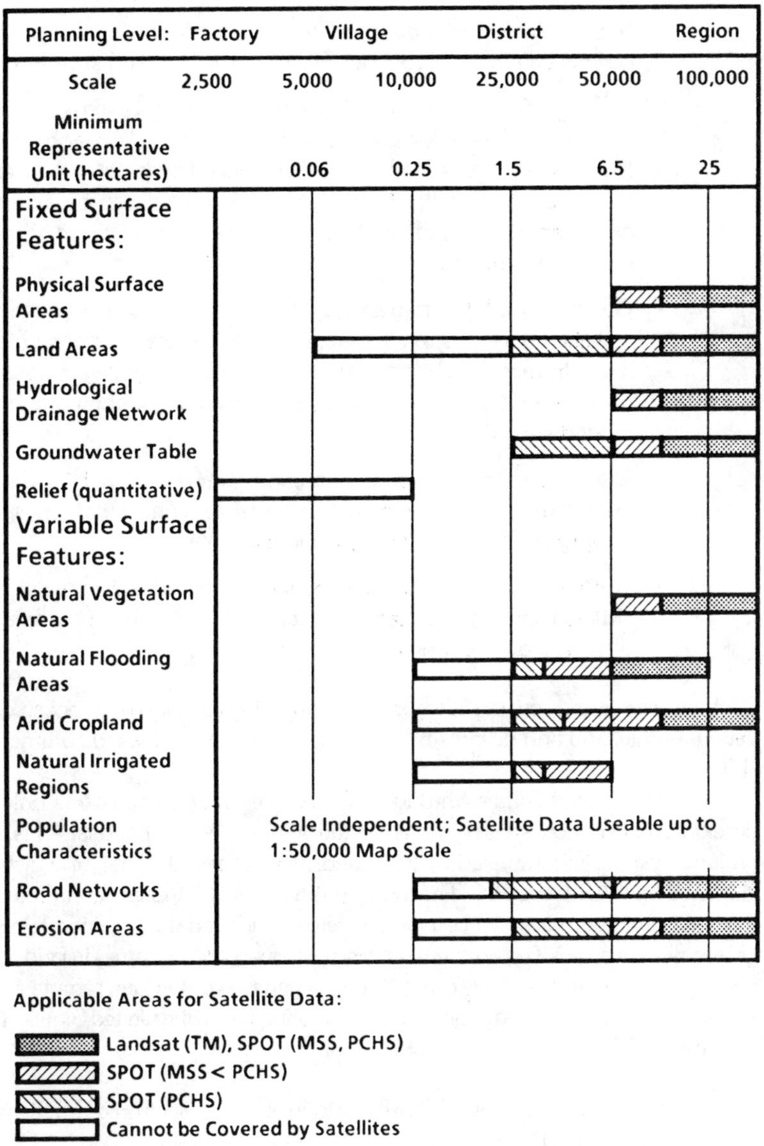

Planning Level:	Factory	Village		District		Region
Scale	2,500	5,000	10,000	25,000	50,000	100,000
Minimum Representative Unit (hectares)		0.06	0.25	1.5	6.5	25

Fixed Surface Features:
- Physical Surface Areas
- Land Areas
- Hydrological Drainage Network
- Groundwater Table
- Relief (quantitative)

Variable Surface Features:
- Natural Vegetation Areas
- Natural Flooding Areas
- Arid Cropland
- Natural Irrigated Regions
- Population Characteristics — Scale Independent; Satellite Data Useable up to 1:50,000 Map Scale
- Road Networks
- Erosion Areas

Applicable Areas for Satellite Data:
- Landsat (TM), SPOT (MSS, PCHS)
- SPOT (MSS < PCHS)
- SPOT (PCHS)
- Cannot be Covered by Satellites

FIGURE 1.
Potential Applications of Satellite Data in the Promotion Phase of a Typical Agricultural Development Project

- **Fishery Applications.** The distribution and abundance of ocean fish in areas of high yield can be delineated, as can polluted waters where the fish and their environment are endangered. Data on environmental parameters such as air pressure, sea surface salinity, and sea subsurface temperature cannot be obtained by satellite measurements; however, automatic ground stations can relay ground-based ocean measurements via satellite.

- **Environmental Protection.** Damage to the environment can be determined, for example, by following the distribution of pollutants in the atmosphere, by identifying large-scale vegetation changes, and by determining the sediment loading of the oceans.

- **Marine Transportation.** Satellite data and analysis of the state of the oceans can be used to reduce ship transit time and save fuel.

- **Geology.** Geological applications of satellite remote sensing are most widely developed for oil exploration. Other applications include location of valuable land and seabed mineral deposits and the extent of degradation of the land. Satellite data can also be applied to geobotanical study of the changes in vegetation indicative of specific geological structures.

- **Geodesy and Cartography.** The most common application of satellite imagery is in cartography. Detailed scale topographic and thematic maps can be constructed economically by integrating satellite data with exact position information from aerial photography and ground survey.

- **City, State, and Regional Planning.** Aerial photographs and stereographic maps, as well as actual computerized enhancement of satellite photographs, permit detailed land-use planning.

- **Archaeology.** Satellite and aerial photographs can help locate structures of early civilizations, even when located in remote areas and obscured by overgrown vegetation.

INTERNATIONAL STATUS OF DEVELOPMENT

Of the three remote sensing categories—meteorological, land, and ocean observations—only land and ocean sensing systems have been applied commercially [1,7].

The United States, through NOAA, presently operates the only fully developed land remote sensing system in the Western World, Landsat. This polar-orbiting system is comprised of five Landsat satellites, the first of which was launched by NASA in 1964. The latest in the series is Landsat-5.[*] In return for a fee, ground stations in 15 countries receive Landsat data.[**] Since 1964, the United States has enjoyed an overwhelming lead over other countries in land remote sensing. Other countries have only recently begun to develop land remote sensing systems based on knowledge and experience gained from the United States. These systems are[***]

- West Germany, MOMS (Modular Optoelectronic Multispectral Scanner)

---

[*]See references [1,2] for a detailed description.

[**]Current fee is U.S. $600,000 [8].

[***]See references [1,2,9] for a detailed description.

- France, SPOT (Systeme Probatoire d'Observation de la Terre)

- India, IRS-1

- Japan, ERS-1 (Earth Resources Satellite)

- Brazil's planned satellite (BRESEX) with a multispectral scanner of 20-meter resolution.

With respect to commercialization issues, the most important fact is that only SPOT was planned as a commercial service from its inception. SPOT is already operational, but its commercial maturity is still unproven. SPOT has been limited to a narrow, high-resolution spectral band. This concept is technically less sophisticated, but it is designed to appeal to a very broad market.

In ocean remote sensing, the United States has an estimated ten-year lead on developments in other countries. U.S. ocean remote sensing satellites include:[*]

- Seasat launched in 1978

- Nimbus-1 through Nimbus-7 launched from 1964 to 1978; Nimbus-7 is still operating.

- Geosat launched in 1985, with special instruments to measure ocean parameters

- TOPEX/Poseidon planned for launch in 1991

- Navy Remote Ocean Sensing System (N-ROSS) scheduled for launch in 1990.

Four non-U.S. systems are noteworthy:[*]

- Japan Marine Observation Satellite (MOS-1) launched in 1987

- European Space Agency Remote Sensing Satellite (ERS-1) scheduled for launch in 1990, with ERS-2 in 1993

- Canada Radarsat to be deployed in the mid-1990s

- Oceanographic Satellites METEOR-2 (1977), Kosmos 1500 (1983), and Kosmos 1602 (1984) of the Soviet Union.

France is participating in the TOPEX/Poseidon system.

The dominant position of the United States in these satellite remote sensing efforts is being eroded by SPOT. It will be further affected in the 1990s by ERS-1, MOS-1, and Japan's ERS, by which Europe and Japan will bridge the technological gap with the United States.

A large market is foreseen for high-resolution stereo images for cartography, geology, and environmental planning. For these applications, highly resolved details of relief features are critical to mapping of the topography [2,10,11].

This brief summary demonstrates the variety of applications of satellite remote sensing data, yet the perception persists that currently available data does not have the resolution necessary for many applications [9]. Since an extensive effort is required to obtain precise high-resolution data, an alternative process—initial mapping from coarse satellite data integrated with data from other sources for detailed mapping—appears to have definite advantages. In any case, a definitive

---

[*]See references [1,2,9] for a detailed description.

assessment of remote sensing capabilities can be made only after satellite remote sensing systems such as SPOT or Landsat have been in commercial operation for an extended period.

DEMAND AND COSTS

The space component of satellite systems consists of launch systems, payload carriers (ELVs, Shuttle), payloads, and ground operations (tracking and control systems). The development of these systems has been carried out by or for the national space agencies (NASA, CNES, DFVLR, and NASDA). The principal motivation for the deployment and operation of the satellites was to meet military needs and/or research objectives. Consequently, the market for space component hardware and services is very limited, strongly influenced by political issues, strongly dependent on government contracts, and distributed among a few nations and private companies.

The user component of satellite systems consists of remote sensors (receiving equipment), ground stations and receivers, and data processing. The development of the market for this component is the main focus of present worldwide commercialization efforts. In fact, marketing of some of these services, such as satellite data transmission, is already based on economic considerations.

In this context, it is important to note that "need" can be defined in two ways: potential need of the data user for processed and enhanced data, and the need of the value-added data processing industry for equipment and systems to meet the needs of the potential users. These needs will, in turn, determine the demand for the systems of the space component.

For the present discussion, the most important single factor is whether the potential market for data and services can support a commercially viable satellite remote sensing enterprise. The relatively small size of the existing market is currently supplied with aerial data from aircraft and balloons. To be competitive in this market, satellite data must have special quality not achievable by aerial photography. Market development and growth must be stimulated through the availability and use of satellite remote sensing data. Parameters such as timeliness, spatial and spectral resolution, and coverage must be very carefully matched to meet the needs of potential customers.

A similar picture emerges for the costs involved in commercial utilization of remote sensing. Table 2 compares typical itemized costs for making maps in

TABLE 2. *Comparative Costs for Making Maps in West Germany*

Parameters	Aircraft		Satellite			
Scale	1:25,000		1:50,000		1:50,000	
Cost per map (DM)	24,000		44,000		26,000	
Area represented per map (km²)	160		640		640	
Costs per square kilometer	DM	%	DM	%	DM	%
Control points	36	24	16	20	—	—
Imaging flight	9	6	7	9	1	3
Analysis	51	34	18	26	9	20
Field comparison	21	14	20	28	25	61
Cartographic processing	33	22	12	17	6	16
TOTAL	150		73		41	

West Germany from aerial and satellite images [10]. The individual cost items for satellite mapping are considerably lower than for aerial mapping, and the cost advantage increases with finer scale. High-resolution satellite images are a major reason for the lower costs of satellite maps. Surprisingly, imaging flight costs are a small fraction (3 to 10 percent) of total map-making costs.

Cartography alone cannot ensure the economic viability of a commercial satellite remote sensing business. Even in countries with the necessary infrastructure (hardware, data processing capabilities), cartography is largely dependent on conventional remote sensing methods and ground-based measurements. Satellite-based methods using Landsat imagery are still employed only to a limited extent. However, the development of higher resolution sensors and the capability for high-speed routine computer manipulation of satellite data will facilitate increased utilization of low-cost satellite mapping [9].

The previous arguments, in principle, also apply to developing countries where information on earth resources is severely needed. These countries can use the remotely sensed data to develop an effective information base. The UNISPACE '82 report notes, "The synoptic view and the possibility of frequent repetitive coverage of large and even inaccessible areas make, for the first time, regional and global monitoring of renewable natural resources and changing environmental phenomena technically feasible and economically attractive" for developing countries [12]. Because they cannot finance these activities, assistance for developing countries would have to be provided through development programs instituted by multilateral organizations such as FAO, UNIDO, and AID, or other national agencies.

Use of satellites in agricultural, forestry, and fisheries development throughout the world varies widely with the production and market structure of individual countries. Individual customers in these resource sectors can be attracted only by providing highly focused and large-scale repetitive continuous coverage. Otherwise the best possible means for market development is to provide specific information and services for groups of users with similar needs and interests. The growing advantages of satellite remote sensing may be more difficult to apply in Europe since its economic systems vary more than the Americas and Asia.

Commercial geological and mining applications do not require continuous observation and real-time data processing. Hence, the petroleum industry is not considered a long-term customer of remotely sensed satellite data.

The unique advantages of satellite remote sensing lie in the opportunity to observe and monitor renewable and nonrenewable resources on Earth and to process and apply the data effectively to resource planning. For this reason alone, the potential need for satellite remote sensing must be considered very high, even though it may not generate sufficient commercial interest in the next few years. Satellite remote sensing systems cannot currently be financially self-sustaining.

With this background on the relevance of satellite remote sensing, the framework conditions for commercial utilization of space for remote sensing will be analyzed at length. For this purpose, the most important factors in the commercialization process will be considered in detail. Also, the existing economic, political, and institutional conditions will be described for each country considered, especially as they relate to a private company contemplating entry into the satellite remote sensing business. Finally, these selected conditions favorable to the commercialization requirements will be compared for their effectiveness in each country and for international comparison of the current status and the short-term outlook for commercialization.

CHAPTER

3
Requirements and Basis for Commercial Utilization

As in any innovative process, commercial space utilization can be considered within a field of favorable and restricting driving forces [13]. Foremost among these are the positive effects of technological developments and market forces that will drive the innovation process, as shown in Figure 2. On the other hand, the development of alternative or competitive technologies, or the presence of seemingly insurmountable barriers to developing a functional operating model of a sound theoretical concept are perceived as negative technological influences. Similarly, unfavorable cost ratios, insufficient market growth, too diverse a customer base with widely varying needs, and limited competition between the existing suppliers are all significant barriers to product innovation. Finally, the envelope of economic and political conditions surrounding technical development can support or restrict innovations; these conditions, in turn, reflect the existing social, cultural, and political systems of individual countries. The interaction of the positive driving forces and the barriers to successful product innovation is very complex and not amenable to the objective analysis required for predicting commercial utilization of space. In the following, therefore, only the needs and demands of potential users will be described in relation to the framework conditions in their respective countries.

FUNDAMENTAL CONSIDERATIONS

All of the preceding applications for satellite remote sensing are of equal interest to *users* in government agencies and private industry. For most applications, timely, continuous, and unlimited availability of data and services are essential for optimum utilization of satellite remote sensing.

From an economic standpoint, the data and images must be available to potential users

REQUIREMENTS FOR COMMERCIAL SATELLITE REMOTE SENSING

- At a reduced cost
- On special order
- On an exclusive basis
- In flexible forms (tapes, maps, tables, and graphs)
- In a reliable manner.

for which the industry must provide suitable guarantees of quality service.

The users place these demands on both the value-added industry as well as on the distributors of raw data, depending on the nature of the products and services required. In turn, the demands of the distributors must be met by the satellite operators to meet the needs of the end users and to operate as a profit-making business. To meet their needs, these *producers*

13

FIGURE 2.
*Schematic Representation of
Innovation Processes*

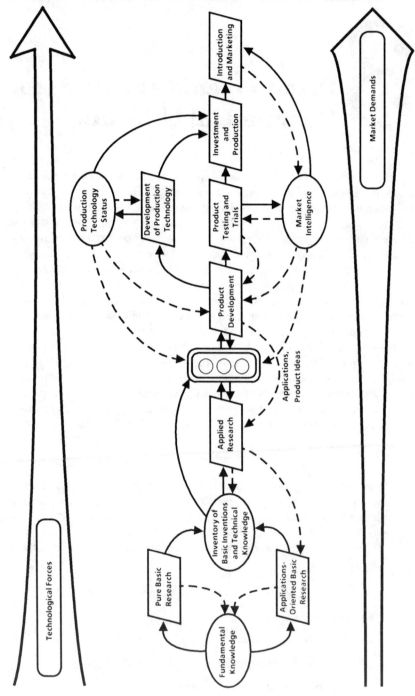

- Will need timely, continuous, and unlimited access to on-line data
- Will demand exclusive rights to access of the data in certain cases
- Will ask for guarantees from the satellite operators.

Clearly, the *distributors* will initiate new business only if

- A sufficiently large market exists
- The monopoly and competitive positions are clearly defined
- Their proprietary rights are protected
- They can obtain cost-effective services from satellite owners.

For the *value-added industry*, other important factors are

- Unrestricted access to other essential information and materials
- Ability to manipulate, analyze, integrate, and make available the spacecraft data to meet the specific needs of the users.

The *satellite operators* can and will meet these demands only if

- The market for satellite services is not controlled by government regulations and restrictions
- Satellite technology has achieved a level at which it is rational and efficient to exchange payloads (such as sensors), service the satellite, permit frequent overflights, and permit redundancy
- Existing and planned sensor technology (resolution and spectral bands) are marketable to the producer segment of the user community.

In addition, the satellite owners will demand uninterrupted and guaranteed launch services from the space transportation industry (ELVs and shuttles). For these interconnected commercial activities to function smoothly, the basic factors below must be favorable to the overall process:

- Legal regulations
- Policy guidelines and political conditions
- Security and military aspects
- Overflight rights
- Access to launch services.

NECESSARY CONDITIONS TO FULFILL THE REQUIREMENTS

Of the framework conditions described in subsequent sections, not all are of equal importance from the perspective of a potential commercial user. Gleaned from our interviews with potential customers of satellite remote sensing about specific user requirements and needs, the framework conditions necessary to develop potentially strong user interest and involvement in commercial space utilization can be identified. Table 3 lists these conditions in relation to each user requirement.

In addition to the overview this listing provides, a number of other related requirements, factors, and conditions affect the commercialization process. Therefore, the following chapters will focus on the basis for international comparison of economic, political, and infrastructural framework conditions for commercial space utilization in remote sensing.

TABLE 3. *Conditions Necessary to Meet Requirements for Remote Sensing*

Requirements	Preconditions
From User:	
Availability	
• Timeliness	High overflight frequency, extensive ground-based infrastructure
• Reliable service	Quality and reliability of operational partners
• Continuous availability	Secured launch availability guaranteed by Government
• Unlimited availability	Politically guaranteed access to data for users; legal clarification of such access
Quality of service	
• Competitive prices	Very dynamic market with many companies; waiver of development costs for system; subsidize costs for application of satellite data
• Customer-oriented service	Service targeted to value-added products
• Flexible purchase conditions	No necessity for purchase of complete hardware units or time slots
• Exclusivity	Proposal for quasi-exclusive value-added products or decision against such policies
From Data Enhancers and Distributors:	
Liability	
• Guarantees	Legal clarification of liability coverage in case of failure of services
Availability	
• Timely access	Same as for users
• Continuous access	Same as for users
• Unlimited availability	Same as for users
Special requirements	
• Guarantees	Eventual Government launch guarantee or launch insurance
• Exclusivity	Political decision on exclusivity
Market development	
• Market potential	Government competition in operation and value addition; Government purchase guarantees for its own use
• Competitive market prices	Exclusion and exemption of development costs
• Monopoly	Positive policy decision for national or regional monopoly within a non-discriminatory framework

16

TABLE 3. *Continued*

Requirements	Preconditions
From Satellite Owners: Requirements for enhancement	
• Copyright guarantee	Legal safeguards of copyright for enhanced products
• Free access to other data (for value-addition)	
• Adequate hardware and technology	If unavailable, should be candidate for Government research support
• Continuity of access to launch systems	Government efforts to develop independent launch systems; to develop joint launch system with other interested countries; or to provide contractual and politically guaranteed access to foreign launch systems
• High level of satellite technology	Support of available technology or development of new Government or private capacities in areas of essential technologies; orientation of space programs for subsequent commercial utilization
Exchange of payloads Service Overflight frequency	
• High level of instrumentation	Support of available technology or development of new Government or private capacities in area of essential technologies; orientation of space programs for subsequent commercial utilization
From Satellite Operators:	
• Free export markets	Abandonment of Government regulation of export markets; exercise of political influence to deregulate these markets on an international level; abandon politically based restrictions
• Guarantees	Assumption of liability based on U.N. liability convention

CHAPTER

4

Economic Considerations for Commercial Utilization

Under economic framework conditions for commercial utilization of space, those aspects will be considered that can influence the decision of a private company for or against space utilization. The key issue in this decision will be the cost/benefit analysis of the space enterprise. In such an analysis all services and support from government agencies are considered positive factors if they result in cost reductions for the private company. These costs must be considered in relation to the social and other benefits in the government's decision to provide financial support for a commercial or profit-oriented space enterprise.

The costs and benefits listed in Table 4 clearly do not include all the factors in the decision of the government or of the private sector to increase involvement in commercial remote sensing ventures. Either relevant information is not available, or the available information is very unreliable, or it cannot be quantified. This listing can begin discussion of the important economic factors and conditions for commercial utilization of space for remote sensing.

Of the cost items listed, actual figures for government spending for space research and utilization are available. The size of this budget need not necessarily correlate with its effectiveness. More important is the location of this budget along the spending curve for the innovation process.

Government initiatives to reduce the economic barriers to space commercialization take various forms [14]. Most directly effective is financial support for applied R&D in specific applications. Support initiatives achieve greater legitimacy if the projects are perceived to have potential social and human benefits. Such projects are often high-risk ventures with long maturation times, factors that make the private sector reluctant to participate. However, a positive decision to invest in the so-called "social good" project is easier to make for the private company if it does not have to assume all of the financial risk. Of course, the project must be in keeping with the company's overall goals and objectives. The use of such initiatives will depend on governmental support that will, at the very least, hamper market-oriented allocation of available economic resources.

This argument also holds for governmental support of industry-initiated technical projects. Indirect financial support for R&D is provided through progressive tax reductions that depend on the size and scope of the R&D activity, even though technical goals may not be clearly defined. Such an initiative provides support for general R&D in space infrastructure and utilization within national and international programs, as well as general governmental support for R&D in the private sector.

Aside from purely financial support, other government initiatives include general information dissemination, organizational and administrative support for user groups, and transfer of scientific knowledge and technology to potential users [14].

As shown in Table 5, the United States' overall R&D budget allocated to space research and applications has continuously decreased since 1970. This has

COSTS AND BENEFITS OF COMMERCIAL SPACE UTILIZATION

TABLE 4. *Costs and Benefits of Space Utilization*

Entry costs for space utilization

For the Government:
R&D expenses for space research in national laboratories and organizations
- Basic funding
- Project funding

Funding of space-related R&D costs for industry
- Nonspecific programs such as support for R&D personnel in general
- Indirect funding of specific programs
- Direct project support

Contributions to multinational space organizations (ESA, U.N.)
Flight costs
- In-house experiments (primarily basic research oriented)
- Industrial experiments (basic and applications oriented)

For Industrial Users:
Outlays for
- Satellites and sensors
- Ground stations
- Data analysis and interpretation equipment
- Sales network

Share of flight costs

Benefits of space utilization[a]

For the Government:
Financial
- No direct payoffs, eventually future tax revenues

Other payoffs
- Support for scientific research
- Dissemination of information of scientific and general interest
- Provide spin-off effects to the community at large
- Establish and support a space hardware industry
- Establishment of favorable initial conditions for commercial utilization by a national industry
- Trade benefits through an enhanced international presence

For Industrial Users:
Financial
- Possible profits in 2–5 years from sale of satellite data products
- Opportunity to market eventual spin-off products

Other payoffs
- Early entry in future advanced areas of information technology
- Stimulation for terrestrial research

[a]For many benefits, it is difficult to provide additional detail.

provided other Western countries the opportunity to catch up in selected areas. The necessary condition for, and a desirable effect of, these efforts was the development of a viable national aerospace industry. As a result, the United States' lead is now based primarily on the size of its annual R&D budget (in absolute terms) and on the size and capabilities of the American aerospace industry. In the foreseeable future, it appears that commercial space activities will not be possible in any of these countries without governmental support for infrastructure and R&D, and organizational, administrative, and financial support from the government, thereby making purely industrial competition impossible. Not only are the ability to develop new markets and the development of competitive products at low cost important, but the size and scope of governmental assistance in the form of financial support, organizational support, and guarantees will determine the commercial viability of a space business enterprise. By means of such direct support, the relevant government institutions also encourage international competition: other countries will be motivated to support market and private development in their own private sectors.

TABLE 5. *Annual R&D Funding for Space Research and Utilization*[a]

	1975	1976	1977	1978	1979	1980	1981	1982	1983	1984	1985	1986	1987
United States	14.5	15.1	12.1	11.3	11.1	9.2	9.2	7.2	5.5	5.2	5.5	5.4	5.95
Japan	7.0	7.7	5.7	6.2	6.0	5.7	5.8	6.1	4.4	4.5	6.8	4.3	4.9
Federal Republic of Germany	4.3	4.5	4.3	4.2	4.2	4.3	4.1	4.2	4.0	3.9	3.9	4.0	4.9
France	5.6	5.4	5.2	5.0	4.6	6.2	4.1	4.3	4.8	5.6	5.0	5.4	6.0
Great Britain	2.2	2.6	2.4	2.1	1.8	1.7	1.9	2.0	1.9	1.9	1.8	1.8	2.7
Italy	8.5	8.8	8.3	9.9	9.5	6.2	5.8	4.1	4.4	6.4	7.8	9.2	9.3

Source: OECD/STIID Databank, July 1988 [15].
[a] As a proportion of total government R&D budgets from 1975 to 1987.

Data on the costs of construction, launch, and operation of satellite remote sensing systems can be gleaned from the experience with the Landsat and SPOT systems. The total cost for a satellite remote sensing system is made up of the following cost components:

- Satellites (including sensors)
- Launches
- Ground stations (control and receiving stations)
- Equipment for processing, analyzing, storing, retrieving, and distributing the data.

The costs for design and manufacture of a complete remote sensing satellite typically range from $60 million to $300 million.

The cost* for launching a satellite on the shuttle in 1985 was $38 million, and in 1986 was set by NASA at $53 million.** Although a U.S. government policy statement of May 16, 1983, declared that after 1988 NASA should charge a "full cost recovery" price [16], NASA has not revised its 1988 price from $53 million [17]. The effect of the Challenger accident on NASA's payload missions has been to ensure that commercial large- and small-capacity ELVs will complement the Shuttle with a mixed fleet for access to space. In fact, a variety of future launch options is being considered, with associated significant potential cost reductions [17,18]. In the policy statement of 1988, the U.S. government stated that after resumption of flight operations, "NASA will provide launch services for commercial and foreign payloads only where those payloads must be man-tended, require the unique capabilities of the STS, or it is determined that launching the payloads on the STS is important for national security or foreign policy purposes. . . ." [19].

The commercial Ariane launch price was initially fixed between $30 million and $35 million per customer for shared launch. The current price (1988) is approximately $42 million [20] for a shared launch on an Ariane-3; compare this with about $50 million for a comparable dedicated Delta launch [20,21]. In fact, commercial launch costs have stabilized at $5,000-6,000/kg in geosynchronous transfer orbit [20]. For smaller satellites (so-called "microsatellites" of up to 50 kg or "minisatellites" of up to several hundred kilograms), rapidly emerging launch options [22] may reduce the cost per kilogram of satellite weight. These small satellites could also be used for earth observation.

Other costs are difficult to calculate. Control and receiving equipment in the ground stations are seldom newly installed; mostly existing (and often government-owned) installations are utilized. The associated costs for use of the equipment can often be reimbursed from profits from future sales. (See Chapter 6 for further discussion.)

Sales are usually the responsibility of the marketing companies and/or agents; the only other costs to be borne are for data storage, analysis and processing, and reproduction. Even for storage and archiving data, which in the United States, for example, is appropriately considered to be a legitimate activity of government interest, the operating company usually does not have to bear all costs for the system.

The greater the governmental financial support and the extent of the research support from the national laboratories, the lower will be the financial investment and corresponding risks for the remaining costs of the space enterprise for private investors and companies. Based on this, the following points can be made with respect to satellite remote sensing:

- The entry costs for satellite data sales, distribution, and value addition are comparable worldwide. The existence of two competing systems reasonably guarantees encouragement and support for the user from the satellite operators. Technical and financial support is available in all of the countries considered for the development of user-oriented data processing systems.

- Establishing a third satellite remote sensing service offers no clear advantages at present.

---

*These prices are charged by NASA for a dedicated payload bay. Since most satellite do not occupy the whole bay, the costs are proportionately lower.

**Since the Challenger accident occurred on January 6, 1986, the price for 1986 is valid only for the single shuttle launch of Challenger.

- In the United States a new service cannot use the Landsat satellites. Hence, the system would have to develop new markets by offering either new technological developments or exclusivity, which is politically unfeasible, or the service will need to offer very high resolution. No governmental financial support is available for these activities. However, the U.S. Department of Commerce is exploring options for a commercial advanced earth remote sensing satellite system beyond Landsat-6 [23]. In fact, Space America, Inc., has expressed interest in constructing a privately financed land remote sensing system as a follow-on to Landsat-6 if the U.S. government decides to discontinue financial support of Landsat [24].

- The interest in an independent national satellite remote sensing system in Japan is principally in ocean observation. Experience gained with MOS-1 may provide a basis for considering commercialization. In fact, Japan is considering the prospect of marketing data commercially [9].

- For political reasons, SPOT will not be followed by another new remote sensing system in Europe. Unlike SPOT, ERS was conceived more as a scientific earth observation satellite, but the experience gained with ERS can form the next step in the commercialization process.

MARKET CONDITIONS FOR THE PRIVATE SECTOR

Space application projects still are not routine in any country in the world. Even so, U.S. telecommunications include sizable private satellite communications networks that offer services for sale. In addition, the Landsat system for satellite land remote sensing has operated since 1972, first by NASA, then by NOAA at the Department of Commerce, and is now being operated privately by EOSAT with government subsidies. Even France's competing commercial system, SPOT, is not a purely private enterprise since the company SPOT Image, SA is in part government-owned.

For the different commercial components of satellite remote sensing systems, the deciding factors for private sector investment and involvement vary sufficiently with corresponding and varying degrees of government involvement. Satellite operators should be able to guarantee service to future customers over a suitable and extended time period (minimum ten years, four satellites) [25]. They must also have the will and capability to install and operate a worldwide data distribution network. Because of these extensive service guarantees and the scope and size of the enterprise, enormous capital expenditures are necessary with a very limited guarantee of any return on investment. Under these circumstances, combining government and private interests is essential to provide some degree of risk insurance and a dynamic market strategy.

In the area of national data distribution and value addition, market potential plays a decisive role for the entry of a private company. The position of the government as an existing or a potential competitor in this activity is also very important. In many countries, the markets for data products and services are limited and the government itself distributes and sells satellite data and images using its own equipment [9]. With respect to commercialization, this situation can best be remedied by limiting the role of government to providing the satellite system and financial support for the venture, possibly within the context of a general development assistance program. Based on the present state of the technology, the overall market for remote sensing products is limited. The applications for remote sensing data mentioned previously indicate that an important factor that will determine market growth is the frequently expressed worldwide interest in satellite data and images [9]. Another factor, particularly for cartography, is image resolution, which must be better than the resolution obtained by other conventional techniques [5]. Ironically, it is the developing countries that have the greatest need for earth resources data from satellites, but these countries often do not have the financial and technical resources to acquire, process, and apply the data [1,5]. In general, potential users will enter the satellite data market only if they can be assured of the future availability of service.

An important question is the number of satellite systems and data distributors that can profitably exist under the most favorable market conditions. Because all remote sensing satellites provide worldwide coverage in their operation, the number of satellites required will depend on their special capabilities—such as atmospheric observation (meteorological satellites) and ocean and land observation (ocean and land remote sensing satellites)—as well as on the frequency of their coverage of individual regions. The SPOT satellite, for example, provides coverage approximately 15 times a year [2]. This frequency is sufficient for all major applications, except for forestry and environmental observations.

The required number of instruments (sensors and cameras) within a data collection system is also important to the commercial viability of satellites and data processors. Given the normal pace of sensor development and the image resolutions to meet most user data needs, as few as three to five competing companies might threaten profitable commercial operation of a satellite remote sensing business. Figure 3 shows the general upward trend since 1978 in overall income from Landsat data sales, although through 1984 this trend was attributable more to price increases,* changes in price structure and policy, and the cost-free distribution of data, than to a significant increase in customer demand [1]. Since 1984 customer demand has increased. However, Landsat sales consisted almost exclusively of raw data (MSS scenes and digital products) sales, of which there is very limited direct use. The U.S. government uses large amounts of raw unenhanced data for which most of the value-added work is done within individual government departments [2,9]. Various industry studies [9] suggest that U.S. and foreign private sector users will spend five to ten times more money on value-added services than on raw unenhanced data from 1987 to 1997. The development of data processing capability is both time consuming and expensive; for example, the oil industry required six to eight years to develop the data analysis and interpretation capabilities in use today.

The relative share of the industrial customer for Landsat data has been fairly constant between 20 and 30 percent since 1973 [9]. However, the U.S. agricultural industry is expected to create a greater demand for remote sensing data, thereby increasing the share of industrial customers. This is because of the very large croplands and the extensive grain trading and brokerage business in the United States. A similar trend in customer demand from the European agribusiness industry with completely different conditions is highly unlikely.

The probable scenario for market development of satellite remote sensing is summarized below.

In the foreseeable future a small number of commercial services will offer satellite and remote sensing data for sale, possibly only EOSAT for Landsat data

---

*In October 1982, NOAA increased the price of data significantly (325 percent for an MSS CCT). Customers purchased more data in anticipation of the price increase, as reflected in the sharp upturn for 1983 in Figure 3.

and SPOT Image for SPOT data. Market growth in this segment will be moderate and occur slowly. A private company could develop intelligent value-added products from corrected spacecraft data and sell the products profitably. The major difficulty facing European companies, and especially companies in Germany, will almost certainly lie in their inability to focus on the worldwide market for such products.

Further, it is expected that SPOT Image will attempt to satisfy customer demands for value-added data and products, leaving only the Landsat satellites as the principal source for raw data. However, even U.S. firms have realized that viable commercial operation of the Landsat system can only be achieved through market growth by offering preprocessed or complete value-added data services to meet user needs [9]. The size of the world market for commercial satellite remote sensing is projected at $0.5 to $2.2 billion in the year 2000, according to a frequently cited 1985 study by the Center for Space Policy (CSP) [26]. This range of values

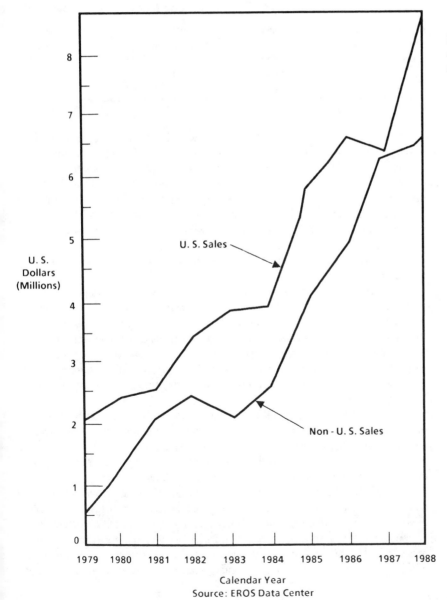

Figure 3.
Income from Sales of Landsat Data by Customer Category

25

was estimated by applying different scenarios and is compared to the 1984 CSP estimate of $2 billion [26]. More recent industry studies [9] predict that cumulative worldwide revenues from the remote sensing market will range from $7.2 to $9.0 billion from 1987 to 1997, with $0.7 to $1.5 billion from raw unenhanced data and $6.5 to $7.5 billion from value-added products and services.

Quantitative values of actual and projected market size for remote sensing data specific to each country cannot be reliably determined. However, qualitative projections based on similar product and user groups provide an overall scenario for market size (Table 6).

TABLE 6. *Summary of Qualitative Market Projections for the Principal Applications of Earth Observation*

Application areas	Actual market size	Potential for application	Potential demand
Agriculture	small	high	high
Fisheries	—	medium	medium
Environment	small	high	medium
Geology	large	high	high
Cartography	small	medium to high	medium

The largest markets for commercial agricultural remote sensing data will remain in the United States in the foreseeable future, followed by Canada and eventually South America. The main reason is the large croplands in these countries. Even for agricultural data of other countries, the United States presumably will be the largest and the most lucrative market because it is the principal agribusiness trader and broker in the world. Demand in Third World countries with large planted fields is also expected to increase, as these countries realize the potential of satellite remote sensing data. One future European application of remote sensing will be the observation and eventual monitoring of cropland areas.

The total market for environmental remote sensing data depends largely on the demand from single or multiple government user agencies. Any market projections must include the future scope of environmental problems as well as the financial capacity of the user agencies. The main customers for such data likely will be the Western industrial nations. Climate and vegetation changes in large cropland areas in developing South American, African, and Asian countries cannot be assessed because these countries are financially unable to invest in remote sensing systems.

The field of commercial geological analysis in the past has been the area of greatest interest to the private sector. However, the market can now be covered largely by presently available data, and no significant growth is expected in the future. The main customers in this market sector are large oil companies and geological exploration firms from the industrial nations.

The market for cartography and land use planning also indicates that countries with large needs normally have limited resources, whereas countries with adequate resources already have satisfied their major remote sensing needs from various other data sources and are not strong customers for remote sensing data. In the United States, however, the development of geographic information systems (GIS) is responsible for increased demand growth. GIS combines remotely sensed data with other types of data (map, census, land use, property ownership) to produce an extensive commercial database [9].

The future potential of ocean observation for fisheries and marine transport is the most difficult to estimate. Even so, some demand exists from financially strong industries in traditional fishing nations with modern fleets.

This situation will not result in a large domestic remote sensing market in Germany such as exists in the United States, so only special market niches can be filled from a broad range of available data products and services to meet user needs in Germany.

For private companies interested in commercial satellite remote sensing, the two major players offering primary data, EOSAT and SPOT Image, have nearly cornered the market. The gap in Landsat coverage between the predicted demise of Landsats-4 and 5 in 1990 [20] and the current launch date of Landsat-6 in July 1991 [27] will be filled by SPOT and Soviet data. New companies offering raw data from new satellites are unlikely to be commercially successful, since the still undeveloped market cannot sustain additional players. * A more feasible scenario is new partners joining with the two existing ventures to offer unique technical and analytical solutions to satisfy customer demands. In such a scenario, financial considerations (including federal subsidies) will not play a major role in private sector involvement in satellite remote sensing. Even so, these maturing developments offer some lessons. New forms of cooperation and coordination between governmental institutions and the private sector have been successful for commercial space activities. This is particularly important for Germany because of its direct competition with two industrial nations, France and Japan (with the exception of the leading space power, the United States), which have traditionally fostered close and flexible cooperation between industry and government. This will be discussed in detail in the next chapter on political framework conditions.

The situation in value addition and sales of satellite data is very different. As noted above, the business potential for private companies does not lie in a large domestic market, making it imperative to develop an international commercial strategy from the very beginning. Successful international competition, in turn, is essential to convincing potential customers of the competitive advantages of remote sensing through data products and services that address their specific needs. However, this strategy should not be limited to the technically challenging high-priced segment of the market, since a large group of less affluent developing countries are potentially important customers, as the previous analysis of market potential has clearly shown.

In addition to governmental support for innovative data processing technology, another important framework condition is governmental support of financially strapped customers from the developing countries. For both these forms of governmental support, the situation in Germany is probably better than that of the other countries considered here, since the government has traditions of significant R&D investment and significant amounts of foreign aid to developing countries.

---

*Soviet, Japanese, Indian, Canadian, and Brazilian satellite systems will be additional sources of primary data by 1995.

27

CHAPTER

5

Legal and Political Issues

In addition to economic framework conditions, political factors exert a strong influence on the commercialization of remote sensing. The obvious differences among the individual countries considered are minimized by international agreements that are unanimously supported by the Western industrial nations.

The international legislation of commercial satellite remote sensing has not yet matured. The United Nations, the Soviet Union, and some developing countries have attempted to extend national laws to cover international rights. These countries have demanded the right to make sensing from their territory and transfer of the data derived therefrom dependent or prior consent from the sensed state [7,28].

This trend towards territorial nationalization of space utilization, in fact, has resulted in the demand of some countries for equitable sharing of geostationary orbits similar to commercial air corridors [29]. Both philosophically and practically, territorial nationalization of space activities is not feasible.

The other option of complete internationalization of outer space activities monopolistically administered by the United Nations will probably require extremely conservative administration with a heavily biased political structure. Such a United Nations management authority will hardly be likely to assign intellectual property rights to commercial users, thereby eliminating a very important criterion for the efficient allocation of resources. This will lead to further erosion of market-oriented elements in the world economy and increase the size of the United Nations' bureaucracy. In addition, such an arrangement would be a disadvantage to countries with significantly advanced space technology. As a result, further developments in this area will be limited [30].

At the present time, the Western industrial nations follow an "open skies" policy, based on the concept of free access to space. According to this policy, prior consent for data and analysis is not required, but the sensed country can participate in remote sensing over its territory based on fair market conditions. This concept is designed to extend and supplement the other historical forums. At the April 1986 meeting of the U.N. Committee on the Peaceful Uses of Outer Space (COPUOS) in Geneva, the "open skies" policy was referred to other United Nations' agencies for ratification [31]; meanwhile, it has been accepted by the U.N. General Assembly [32].

The concept of free access also agrees with the Western understanding of the Outer Space Treaty of 1967 (Germany is also a party to the Treaty), particularly Article IX, which states that the peaceful use of space for the benefit of humankind should not be restricted [33]. Thus, commercial activities in space are certainly not excluded (compare Article 19 of the U.N. Declaration of Human Rights, guaranteeing the free flow of information). This international treaty also forms the basis for the United States' remote sensing policy, which has made available civil remote sensing data from its own satellites to all interested users on

NATIONAL AND INTERNATIONAL LEGAL PROBLEMS

a nondiscriminatory basis. The different prices charged by SPOT Image for standard format data and for special attributes (*e.g.,* cloud-free scenes) were considered by the United States to be potentially discriminating [1]. However, SPOT Image has very clearly enunciated its policy of nondiscriminatory remote sensing data sales (the higher price charged for special scenes is needed to recover the cost for additional effort in reprogramming the sensors, although these special-attribute scenes are also made available for general sale) [34].

A joint venture proposed by MBB-ERNO, COMSAT, and Stenbeck Reassurance Co., Inc. (SPARX Corporation), for exclusive use of a MOMS system on the Space Shuttle, with U.S. companies exclusively marketing the data, was rejected by the U.S. government because it was in direct conflict with the existing "open skies" policy [1].

This "open skies" policy has been useful in blunting the criticism of other U.N. member nations against the Landsat satellites. The United States has always argued that remote sensing is a peaceful use of space in which the national sovereignty constraints have no valid application [1,7]. Given the current low image resolution of tens of meters from civilian satellites (military surveillance satellites have much higher resolution—under 1 meter), no major difficulty is expected with this policy (although eventual problems may arise with higher resolution of less than ten meters, such as with the MOMS system).

A project such as MOMS might, in fact, be opposed by other U.N. member states for national security reasons, even if the "open skies" policy were to be canceled. At the present time, it is the declared political intention of the Western industrial nations not to revise U.N. guidelines for space utilization in favor of the positions of the Soviet Union and the other developing countries, but rather to abandon U.N. guidelines completely. In fact, the Western countries would even tolerate the dissolution of COPUOS. Recently the United States has begun to express reservations about continuing its "open skies" policy because of continuing improvements in the resolution capability of civilian remote sensing satellites. However, the self-imposed restrictions in the 1970s on high image resolution of civilian satellite sensors [35] have been abandoned by the United States in the face of growing international competition.

Privatization of outer space activities basically involves compliance with the 1967 Outer Space Treaty, which makes the individual states responsible for meeting the treaty obligations (Articles VI and IX) [33]. These are also the operating conditions for a commercial enterprise with governmental control. International control mechanisms will be needed if a multinational consortium is established.

The launching state is similarly responsible, according to the Liability Convention, for personal and property damage caused by space objects launched from its territory or if the launch was procured by the state [36]. This convention covers only material damage from space activities; nonmaterial damage such as commercial failures from space utilization are not covered.

With respect to copyright of remote sensing data, no international regulations exist as yet. However, the individual regulations of the distributing organizations for satellite data, which differentiate between raw data and enhanced data, are under discussion; copyright will be claimed only for unenhanced raw data. On the other hand, SPOT Image assesses variable royalty payments for transfer of unenhanced and enhanced data to third parties [34].

The unauthorized use of copyrighted remote sensing data will proliferate irrespective of the nature of the copyright regulations. Copying of computer tapes and photographs is widely prevalent in public and private companies. This practice can hardly be controlled by copyright laws, as the unauthorized transfer of computer software has shown [7].

30

The above discussion has pointed out a number of legal problems in commercial space utilization. These problems can be solved through appropriate national or international agreements. However, the political will to reach such agreements also depends on the commercial interests and priorities of the individual countries.

For satellite remote sensing, the overall situation regarding commercialization is simpler than that for microgravity utilization. The latter is probably contingent on the joint international efforts to construct Space Station Freedom in the next ten years [37]. International legal problems will not arise in remote sensing if an effective "open skies" policy is supported by all nations. The formal principles of the "open skies" policy will remain basically applicable in the foreseeable future. However, some commercially important market sectors can be developed only from semi-exclusive information obtained from appropriate value addition to unenhanced satellite data.

The legal and political considerations presented in this section can be summarized as follows:

- Federal financial support for satellite remote sensing systems is preferred by companies in those countries (outside the United States) with flexible cooperation between the government and the private sector.

- The limit of 10 meters for ground resolution of space-borne sensors, for reasons of national security, will survive at least as long as there are only two major providers of unenhanced raw data.

- Eventually the developing countries will find it in their interest to stop opposing the "open skies" policy, so the value of government guarantees of access to space activities will diminish with time.

- The success of commercialization depends more strongly on trends in customer demand than on unique national factors regulating the products and services offered.

POLITICAL STRATEGIES AND OPTIONS

The national U.S. philosophy of strict organizational separation of government and private sector activities creates major difficulties in defining the nature of, and the conditions for, the privitization and commercialization of government-run programs. This is particularly true in satellite remote sensing: the government has a certain political interest, but the economic advantages lie in transferring the system to the private sector. Various options have been suggested [35]:

- A government-owned and operated system that would sell services to the private user under conditions determined by the government

- A government-owned system operated for the government by a private contractor, but which allows the contractor to develop its own marketing approach to private customers, within well-defined limits

- A privately-owned and operated system supported by the government with definite purchase guarantees and limited exclusive rights, but the private owner is responsible for capturing the private sector market

- A fully commercial system, possibly even with competing companies, with an entirely market-based pricing policy that applies even to government customers.

Government political support for private sector ventures is often essential to successfully overcoming the eventual international opposition to such ventures.

A further problem arises when organizations are formed in addition to existing federal facilities since their activities will overlap in at least the initial phases. This situation would become awkward when private sector companies begin competing with federally subsidized companies [1].

In France, the separation of the public and private sectors traditionally has been less important. State-owned and private companies often are full and equal partners in an economic business venture, even in the international arena. Federal subsidies to private companies are not the exception to open up and penetrate foreign markets. The role of the German government is probably more similar to that of the United States than France.

The French model, Groupement d'Intéret Economique (GIE), was adopted by the Commission of European Communities (CEC) through a business ordinance of the Council on July 25, 1985, for the creation of a European Economic Interest Association (EWIV) from July 1, 1989 [38]. Within this association, companies and individuals from the private and public sectors of the CEC member states can meet together to strategize for improving overall economic performance in wide-ranging areas of members' interest. Each member of the association pays its own costs and expenses; only the costs of association activities are distributed to the members based on a pre-agreed formula. The EWIV can be used, for example, to support the association through federal subsidies in the high-risk startup phase in situations where the government share is disproportionately large.

By contrast, Japan combines all of its national capabilities under government direction in the initial stage of development of a new technology; this eases its entry into the free market once the technology has reached commercial maturity. The variety of existing relations between the government and the private sector leaves no clear separation between their spheres of activities.

The different initial (economic) positions of the countries interested in space commercialization has led to different approaches—a situation that will continue in the foreseeable future. This means that the European countries will continue to choose a multiplex strategy to balance the advantages and disadvantages of cooperation within the ESA. This strategy includes: bilateral or multilateral optional ESA projects; collaboration with NASA; future collaboration with Japan and other space-active countries; and, in individual instances, cooperation with the Soviet Union. This form of cooperation gives the European countries sufficient balance of emphasis between federal agencies and private companies, yet maintains adequate information and technology transfer. This is discussed in greater detail in Chapter 6.

For Germany, this strategy of real division of labor could mean, for example, that in the future no new major R&D efforts will be undertaken in satellite remote sensing. Rather, the emphasis will be more on investigating future applications of previously developed sensors and a further development of data processing techniques. With its leading European position in satellite remote sensing and in launch systems, France can be expected to make stronger efforts to capture a larger share of the commercial space market. In contrast to Great Britain, which has only recently organized and consolidated its space activities, Italy already has started to develop detailed plans and specific milestones for its participation in Space Station Freedom. In Japan, future emphasis will be on further development of an indigenous launch capability. The United States, with its much larger research budgets, is not nearly so limited in the choice of R&D fields; even so, it appears that NASA also will have to prioritize, as well as develop closer international cooperation.

The following recent and ongoing developments have radically changed the market for launch services [9,20,22]:

- The successful entry of the Ariane series of vehicles in the satellite launch business as the first commercial competition for the U.S. Space Shuttle

- The Soviet Union's Proton vehicles, which are capable of launching all existing and planned satellites

- The Chinese contracts to launch commercial satellites aboard its Long March rockets

- The successful test of the three-stage version of the H-I vehicle in August 1987, which will be replaced in the early 1990s by the completely Japanese designed and built H-II vehicle with a capability of lifting approximately 8,000 kg to LEO and approximately 3,860 kg to GTO

- The activities of U.S. manufacturers who have developed large-capacity, highly reliable ELVs such as Atlas-Centaur, Titan, and Delta

- The emerging market for commercial small-capacity ELVs in the United States, Europe, and Japan.

The overall trend is toward a segmented market in the 1990s. The continued existence of such a fragmented structure of launch providers can, however, be ensured only by a corresponding demand for launch services primarily from the telecommunications industry. Expert opinions differ widely on future demand for launch services. Optimistic forecasters estimate approximately 15 to 20 civilian satellites launched per year in the next decade, whereas pessimistic forecasts project only a slow increase in the demand: 10 to 12 launches per year through 2000 [39]. The actual number of rocket launches per year will be reduced since Ariane and Titan are frequently dual launches. NASA's official manifest [27] of approved and requested payloads lists 19 large-capacity ELV flights from 1989 to 1994, which is less than even the pessimistic forecast.

U.S. space transportation policy [17] has recently moved toward strict separation between launches of public (NASA, DOD, ESA, and other foreign countries) and private sector payloads. Military and scientific payloads (including microgravity experiments), to be launched in the early 1990s on the four-orbiter Shuttle fleet, will be handled differently from private or semiprivate launch services on ELVs by a revitalized private industry. This has greatly facilitated private companies' involvement in providing launch services. In the Commercial Space Launch Act of 1984 (Public Law 98-575) [40], the U.S. government fully endorsed and supported the private sector in commercial ELV investment, but at the same time offered to launch commercial satellites at subsidized prices. With the stepwise increase in Shuttle prices to higher levels than ELVs, to reflect its advantages over the ELVs (manned and reusable *vs.* unmanned and expendable), the actual abandonment of commercial satellite launches on the Shuttle (except for Shuttle-unique payloads) represents an important step in official U.S. policy to largely commercialize the market for satellite launch services. In view of the above mentioned additional launch service providers for services available now and in the 1990s, further political efforts to ensure access to launch systems are hardly required. For users interested in commercial remote sensing, access to launch systems will no longer be a problematic issue.

The conclusion to be drawn from this discussion is that legal and political framework conditions will assume less significance for the private operation of a satellite remote sensing system. All interested users of such a system in the 1990s can choose from a wide variety of launch systems, if the user is willing to operate within the previously described international regulations of the "open skies" policy and has self-imposed restrictions for use of sensitive data from increasingly higher resolution sensors (which could have national security implications). For obvious reasons, it is easier to reach the necessary legal, financial, and contractual agreements for launch services within the user's own country than with a foreign country. Access to a national launch system, therefore, is always an advantage for the customer. This situation currently exists in the United States and the Soviet Union, as well as ESA members, Japan, and China. Even though Ariane is a coordinated ESA program, France leads the effort, manufacturing the major components and owning and operating the Kourou spaceport in French Guyana. The continuing positive dynamism of European space policy will almost certainly give other ESA member states, especially Germany, access to an "indigenous" launch system.

CHAPTER

6 Organizational and Institutional Infrastructure

If it is assumed that the preceding political framework conditions change slowly, if at all, in keeping with national interests and strengths, and consequently are of limited value in fulfilling the conditions for commercial satellite remote sensing, then it becomes even more significant to discuss how the existing national organizational and institutional infrastructure encourages the involvement of the private sector in remote sensing enterprises.

The overall situation of each country considered is described below.

WEST GERMANY

The overall goals of the German space program are [41,42]:

- To apply the potential of space technology for basic research, innovation, and strengthening the competitiveness of industry in cooperation with other European countries and the United States

- To promote international cooperation.

Institutions

The Ministry of Research and Technology (BMFT) and the Ministry of Transportation are responsible for putting the above objectives into practice in the remote sensing segment. These agencies oversee large research establishments such as the German Aerospace Research Establishment (DFVLR) and the German Weather Service (DWD). Other organizations within the telecommunications (PTT) segment and the Ministry of Defense are also involved.

Remote sensing is fourth in order of funding support of eight research fields pursued by DFVLR [43]. The planned budget for 1988 for remote sensing was approximately DM38 million [44]. The space systems research field also includes some remote sensing projects. Regarding remote sensing systems, the official DFVLR document states:

> Research efforts are primarily concentrated on the advanced development of remote sensing techniques and systems in the optical, infrared and microwave ranges. DFVLR has also developed simulation and interpretation models for processing the information obtained. . . .
>
> Satellite data are received, stored, archived, preprocessed and processed by subject matter. . . .

Further activities are centered on the planning and project management for developing satellite sensors as part of DFVLR's involvement in international space projects [43].

Germany, unlike France and Great Britain, does not have a national space authority.

Activities

The German Remote Sensing Data Center (DFD), within the DFVLR Satellite Operations Center in Oberpfaffenhofen, had a total budget of DM 6.8 million in 1986, of which approximately 59 percent was earmarked for the acquisition and sale of data [43]. DFD was established to acquire and provide remote sensing data and to process, enhance, analyze, and interpret the primary data from space. The latter functions are its principal research activities in support of basic research. The technical and scientific services of DFD are available to users from government, industry, and the scientific community at nominal cost. A transfer of part of DFD services to the private sector is under negotiation [44].

The conditions under which DFD can transfer Landsat data to third parties are similar to those specified to date by NOAA [45]: "Publication in newspapers, journals, books or scientific reports is permitted if the source is listed. DFVLR should be informed prior to each publication (copy of preprint). Resale of photographic reproductions is not permitted . . . these data are provided on the condition that they will not be resold as whole or in part, and will not be reproduced for sale."

These terms and conditions are not favorable to commercial users. However, the right to manipulate and enhance the data for commercial purposes (value-added industry) is not affected by these conditions [45].

DFD also distributes SPOT and Landsat data. The conditions specified by SPOT Image are described in more detail in the section on France.

Research

Government-supported research in remote sensing is concentrated in the areas of atmospheric physics and climate, meteorology, cartography and land utilization, ocean and ice research, and geodesy. German research programs for commercial utilization of space in remote sensing have contributed to the development of more efficient satellite components and sensors. Good examples are [42,46]

- The development of the metric camera, which was provided to ESA as the German contribution to the Spacelab-1 mission. The instrument was developed with approximately DM 8 million support from BMFT over an eight-year period with matching levels of support from DFVLR.

- Data with a spatial resolution of 25 meters are archived at DFVLR. Photographic images are available from the archive for a nominal fee. If shuttle launch costs are included, the prices for remotely sensed photographic images from Landsat-TM and SPOT would be comparable.

- MOMS is a BMFT-funded project to develop an electronic spectral imaging instrument with a ground resolution of 20 meters. About 85 percent of BMFT funding was for equipment development and about 15 percent for data evaluation. The instrument was successfully tested on Shuttle flights STS-7 (June 1983) and STS-11 (February 1984).

- A consortium of MBB-ERNO, Stenbeck Reassurance, and SPARX Corporation was prepared to market selected 20-meter resolution, high-quality data to the commercial user community. The plans were dropped, however, when NASA pointed out that the "open skies" policy is not compatible with such an exclusive use of the data.

There are plans to develop additional instruments such as [46]:

- PRARE, a microwave positioner to increase positioning accuracy to 10 cm, to be flown for the first time in 1990 on the European ERS-1 satellite

- MIPAS, a Michelson interferometer, to sense trace gases in the stratosphere

- X-SAR, a synthetic aperture radar instrument developed jointly with Italy (PSN/CNR) to be flown on a joint mission with NASA

- LIDAR, a laser system for detection of trace gases in the atmosphere.

Budget

Germany finances its satellite remote sensing projects primarily within the framework of ESA activities. In the 1987 federal budget of DM 1.05 billion, DM 550 million was earmarked as the mandatory German contribution to ESA [47]. Contributions to the optional ESA earth observation programs are nearly 23 percent [48]. By contrast, federal expenditures for national space programs in Germany were DM 505 million in 1987, of which nearly 2 percent was budgeted for instrument development, experiments, and systematic investigation of remote sensing applications [44]. In 1986, nearly 70 percent of the budget for remote sensing was earmarked for equipment development, and 30 percent was used for experiments and data evaluation.

For 1988, the planned expenditure of DM 1.2 billion represents a significant increase from the space budget of 1987. By 1991 the total space budget will be DM 1.559 billion, a 28 percent increase over 1988.

Private Sector

In Germany, no private institution has yet been established for the sale and distribution of satellite images. Several companies, such as Gesellschaft für angewandte Fernerkundung (GAF) and Prakla-Seismos AG [8], are involved in adding value to remote sensing data, but these companies mainly depend on aerial photographs and use satellite photographs to a limited extent. The reasons cited for this situation are the existing difficulties in procuring satellite data, as well as inadequate ground resolution of the satellite photographs.

Satellite data are required mostly in the initial stage of the planning process. Users will forego the use of the data if not delivered in a timely, continuous manner. For more detailed planning, however, aerial photography and ground survey are still required.

Comments

Germany possesses, at least quantitatively, an infrastructure adequate for satellite remote sensing. This has enabled it to develop outstanding sensors and has given active researchers and institutions access to bilateral and multilateral cooperative projects. However, the marketing and sale of Landsat and SPOT data by a federal agency (DFD) shows that the potential for application of remote sensing data still has not led to a truly private user market for this data. This cannot be attributed only to the lack of knowledge and awareness of private industry. More to the point, if as suspected federal agencies cannot increase the market for data products and services, private industry's skills in marketing and innovation should be employed more extensively than before to increase the overall market. Since federal support for evaluation and interpretation technology has already given promising results, attempts should be made to expedite the transfer of these results to industry for commercial utilization. Support for such technologies should emphasize rapidly usable solutions and should be increased for private sector R&D projects.

FRANCE

France bases its scientific space program on a policy of cooperation [49]. In commercial utilization the French government will act as the pacesetter to demonstrate the potential of space for commercial applications.

Institutions

The responsibility for the French space program lies with the Ministries of Industry and Research and of Industrial Reconstruction and Foreign Trade. The important tasks of both Ministries is long-term program development and preparation for authorization of the annual budget for the Centre National d'Études Spatiales (CNES). The budget is proposed by CNES and is generally funded through the Ministry of Industry and Research.

Within Western Europe, France has the largest and most expensive national space program. CNES is the national agency that manages most of the French space program and dispenses the space budget.

Unlike DFVLR, CNES operates only research facilities that are essential to carry out its own major activities (*e.g.,* quality assurance laboratories). Present CNES activities are still dominated by the "classical" areas—launch vehicles (development and limited operation of Ariane), telecommunications, and remote sensing—although funding for microgravity utilization is increasing.

CNES has also initiated formation of important commercially oriented ventures, particularly in remote sensing. Other specific activities, such as the marketing of civilian space products from French industry and opening of new markets, have been excluded from CNES authority. To carry out these activities, as early as 1974 a Groupement d'Intérét Economique (GIE) Prospace was founded. The 52 members (1987) of Prospace are from CNES, aerospace and satellite manufacturers, the telecommunications industry, and several banks [50].

The Groupement d'Intérét Economique pour le Développement de la Télédéction Aérospatiale (GDTA) is similarly organized. This organization includes, in addition to CNES, the following federal agencies as members:

- Institut Français du Pétrole (IFP)

- Bureau de Recherches Géologiques et Minières (BRGM)

- Bureau pour le Développement de la Production Agricole (BDPA)

- Institut Géographique National (IGN).

GDTA functions include education and training in remote sensing, provision of quality services (*e.g.,* distribution of Landsat data through the ESA

Earthnet system, operation of simulation flights for SPOT), and marketing support activities [51]. An independent company, Collecte Localisation Satellites (CLS), was established in early 1986 for commercialization of the ARGOS system (described below) [47].

CNES also cooperates closely with private industry, universities and academic organizations, and government laboratories for basic research, such as Centre National de la Recherche Scientifique (CNRS).

In 1984 a cooperation agreement was signed between CNES and BRGM for the development of remote sensing applications for geological and mineral exploration [51].

A company called Scot was created in 1986 to provide consulting services in remote sensing [47].

Activities

In connection with remote sensing, the French company SPOT Image, S.A., founded by CNES, must be mentioned. Like Arianespace (discussed under European Space Agency), CNES is the largest shareholder in SPOT Image with a controlling block of shares (greater than one-third of capital, according to French law) [52]. Thus, SPOT Image is a semiprivate corporation with a commercial structure. The shareholders, with their shares in the corporation, are listed in Table 7.

The goal of SPOT Image is to open new markets for satellite data on a much larger scale than is possible by a completely government-owned organization. The starting capital of FFr 25 million was increased by FFr 24 million in 1984 [34].

TABLE 7. *Shareholders and Their Percent Shares in SPOT Image*

Shareholders	Percent share
French:	
CNES (Centre National d'Études Spatiales)	39.0
BRGM (Bureau de Recherches Géologiques et Minières)	10.0
IFP (Institut Français du Pétrole)	10.0
IGN (Institut Géographique National)	10.0
Matra	7.5
SEP (Société Européenne de Propulsion)	7.5
IRDI (Institut Régional de Développement Industriel Midi-Pyrénées)	1.2
Banque Nationale de Paris	1.2
Valorind (Société Générale)	1.2
Foreign:	
SSC (Swedish Space Corporation)	6.0
Etat Belge (Service de la Programmation de la Politique Scientifique)	4.4
Others:	
Belfotop Eurosense S.A. and Walphot S.A.	
BTMC: Bell Telephone Manufacturing Company	2.0
ETCA: Études Techniques et Constructions Aérospatiales	———
TOTAL	100.0

The French earth observation program SPOT (Système Probatoire d'Observation de la Terre) was started in 1978 in cooperation with Belgium and

Sweden. The SPOT-1 satellite was launched in February 1986 and has returned over 750,000 images to Earth [34]; SPOT-2 (scheduled for launch in 1990) and SPOT-3 will ensure the continuity of satellite service through 1990; additional satellites (SPOT-4 and possibly SPOT-5) will provide services for the next ten years. In June 1985, a commitment was made to fund SPOT-3 and SPOT-4. SPOT-3 is under construction, and SPOT-4 is in development. The prime contractor for the manufacture of the satellites is Matra, together with Aérospatiale and Société Européenne de Propulsion (SEP) as subcontractor. Special agreements exist with Sweden and Canada for receiving, evaluating, and the marketing rights for SPOT data. SPOT Image has established its own ground station in Toulouse, France. Other earth stations are Prince Albert in Canada; Reston, Virginia, U.S.A.; Hyderabad, India (started in May 1987, but shut down from May to August because of excessive cloud cover); and the ESA station in Maspalomas, Spain (started in November 1987). Additional earth stations commissioned in 1988 or to come on line shortly thereafter are in Cuiaba, Brazil; Riyadh, Saudi Arabia; Islamabad, Pakistan; Dhaka, Bangladesh; Bangkok, Thailand; Beijing, China; Katsura, Japan; Quito, Ecuador; and Nairobi, Kenya. The satellite data have a ground resolution of 10 meters (panchromatic mode) to 20 meters (multispectral mode); the lower limit of altitude resolution in pseudostereo format is 5 meters. The area imaged in each scene is 60 km × 60 km.

SPOT-1 follows a polar orbit at an altitude of 832 km, which takes it over the same spot on Earth at the same time of day every 26 days. Specific areas can be imaged from different angles at intervals of two to five days. Further improvements are planned for later satellites, such as placing other instruments in addition to the optical sensors on board the satellite (*e.g.,* a precise positioning instrument on SPOT-2). For SPOT-3 and SPOT-4, the planned improvements include an additional sensor in the middle infrared range for measuring moisture in vegetation, a new radiometer for monitoring vegetation and oceans, improved panels for the solar arrays, and larger capacity batteries.

Since its inception, SPOT Image has carried out a wide spectrum of activities to develop a marketing and sales organization for SPOT data. In 1984, contractual agreements were reached with 25 countries for distribution of SPOT data, including the United States, Germany, Poland, and Hungary. Special agreements have also been reached with Sweden and Canada:

- **Sweden.** CNES and SPOT Image are shareholders in the Swedish company Satimage, a subsidiary of Swedish Space Corporation. The company was founded to market SPOT data received in Kiruna in the Scandinavian countries.

- **Canada.** In 1984, SPOT Image signed two agreements with Canadien de Télédétection. One regulates the receipt and sale of SPOT data in Canada; the other agreement regulates the receipt of data over U.S. territory and the transfer of this data to SPOT Image Corporation, a U.S. subsidiary of SPOT Image.

For evaluation, cataloging, and archiving of SPOT data, SPOT Image has developed suitable mechanisms for cooperative projects with appropriate institutions in 1984.

SPOT Image is expected to be self-sustaining from the income from sales of remote sensing data from SPOT-1 (sales in 1987 reached FFr 61 million and FFr 80 million in 1988) and later SPOT-2, but it has received financial compensation for the delayed launch of SPOT-1. Later, SPOT Image should also be able to finance the launch of follow-on satellites. In that case, SPOT Image will bear the costs for the basic operational payload and its launch, all other advanced developments of the system will be (at least in the development phase) financed by CNES. Meanwhile, CNES is expecting to realize significant cost savings by combining the manufacture of SPOT-3 and SPOT-4 with the military surveillance satellite Helios.

Research

Scientific experiments in space research (as in Germany) are carried out on various missions, including joint missions with NASA. France also cooperates with the Soviet Union. As in Germany, France manufactures individual components developed for space applications, as well as complete satellite systems such as SPOT.

Along with the SPOT system, a program to develop applications for SPOT data was also introduced: Programme d'Evaluation Preliminaire des données Spot (PEPS) [34]. Within this program, SPOT data were provided to 130 selected participants on special terms. The technical review of the resulting applications ideas was performed within CNES with assistance from an international committee consisting of three scientists from France and one each from Brazil, Belgium, Thailand, Holland, Senegal, Canada, the United States, Japan, and Sweden. Another very successful French-led development is the ARGOS data collection system (DCS), which is used with polar-orbiting U.S. satellites. The DCS collects signals from stationary and drifting terrestrial relay stations that monitor local environmental data not obtainable by satellite systems alone. The special feature of ARGOS is its ability to locate drifting platforms (within 2 to 5 km). The data are collected in the United States and relayed to France. CNES is responsible for data processing, system operation, and user contact. CLS was formed as a GIE in 1986 to market the ARGOS system.

Since the ESA launch facilities are located on French-owned territory and France has led the development of the Ariane launch vehicle, of all ESA member states, France is in the best position to achieve autonomy in satellite launches.

Budget

The French space budget in 1987 was FFr 5.87 billion, including FFr 2.3 billion in payments to ESA [47,53]. This leaves approximately 61 percent of the funds for use within France or for bilateral projects with countries outside ESA [53]. This is very different from other Western European countries, who spend a much larger percentage of their total space budgets within ESA. Clearly France places much greater emphasis on its national space programs.

CNES, as an Etablissement Public Industriel et Commercial (EPIC), is supported nearly 80 percent by the government; 20 percent of its budget is covered by income from such commercial activities as ground control services, operation of the launch facilities, and laboratory and management services.

Private Sector

The private sector plays a limited role in commercializing satellite remote sensing in France. The semipublic SPOT Image includes the large French aerospace firms and some commercial banks with smaller shares (Table 7). On the other hand, SPOT Image is expected to operate on a strictly commercial basis, which essentially requires greater private investment. If SPOT Image increases its efforts in data enhancement, competing private firms will be largely excluded from the market.

41

Comments

The situation in France illustrates its unique approach to achieve its current commercial success in space activities. The framework conditions for the individual phases and segments of this approach have been

- Strong conception and definition authority vested in a single space agency

- Market-oriented product definition

- Availability of sufficient financial resources

- A mature and productive domestic space industry despite purchase of major system components from abroad (including the United States, Germany, and Great Britain)

- Early tie-ups between interested government and private institutions (e.g., GDTA)

- Semipublic corporate model, which combines the advantages of federal guarantees with private sector mobility and dynamism.

Still not assured, however, is the further involvement of the private sector, the long-term profitability of the ventures, and the development of a large supplier and user community in France. Hence, despite its notable initial successes, the French semipublic enterprise model probably does not represent a long-term solution to self-supporting private commercialization.

ITALY

The goal of the Italian space program is to position Italian industry to more actively participate in space activities. Efforts toward commercial utilization have to date been concentrated mainly in the telecommunication segment [54].

Institutions

The national Italian space program, in accordance with the resolution of the Interministry Committee for Economic Planning (CIPE), is centrally coordinated and managed by the National Research Council (CNR). The budget of CNR is proposed by the Ministry for Scientific and Technological Research and included in the national budget. Following parliamentary approval, the funds are released to CNR by the Finance Minister. Since Italy also participates in international space programs, the Foreign Minister must approve the transfer of Italian contributions to ESA.

In 1979, CNR established within its organization a national space administration, Piano Spaziale Nazionale (PSN), to manage the national space program. PSN is responsible for defining the scope of space programs and for organizing the activities necessary to carry out the programs, including awarding industrial contracts and monitoring contractor work. Financial resources are provided by CNR; PSN provides the technical expertise [55].

An independent Italian space authority (ASI) with its own budget was established in 1988 to replace PSN. This new national agency will be responsible for all Italian space activities.

Italy also cooperates with ESA activities since its national program covers only preoperational satellite systems; the actual market, however, is perceived to be in equipment for operational systems.

A special characteristic of Italy is that the major aerospace companies (with the exception of the Fiat-owned Snia BPD) are mostly owned by the Italian government. Under the umbrella of a government holding company, the Institute for Industrial Reconstruction (IRI), the companies are organized in two groups: Finmeccanica, which also includes Aeritalia; and the telephone financing company, Società Finanziana Telefonica p.a. (STET), including Selenia-Spazio and Telespazio,* which deal exclusively with space products [56].

Activities

Prospective customers for remote sensing presently include only government agencies such as the Interior Ministry, Ministry of Agriculture and Forest, and the Merchant Navy. Thus, Italian space activities concentrate on developing and improving systems to evaluate remote sensing data for government functions.

Research

A much larger program is concerned with evaluation of satellite data for crop forecasting. The program is being conducted by a Consortium of Telespazio, Italeco, and Aquater under contract to the Agriculture Ministry. It was planned to be tested for the first time in 1986 in Sardinia.

Budget

The annual Italian space budget was about L 115 billion in the early 1980s, increasing to L 362 billion for 1985 and L 698 billion in 1987 [47]. In the coming years, the budget is expected to increase to L 700 billion annually, especially because of the European Columbus Attached Laboratory for Space Station Freedom, which is increasingly becoming the focal point of Italian space activities. Remote sensing takes about 15 percent of the total budget; microgravity utilization, 5 percent; and Columbus, 25 percent of the total budget [48]. The remaining 42 percent is used mainly for the telecommunication segment and for ESA contributions.

Private Sector

Remote sensing activities in Italy are conducted primarily by the government-owned Telespazio. Any other private sector activities are not known.

Comments

Satellite remote sensing is not a major focus of Italian space policy. Commercialization in Italy is centered within the framework of the traditional semipublic industrial structure. Whether an appropriate restructuring policy (transfer to private sector) will produce far-reaching changes in the existing framework conditions is difficult to evaluate. Semiprivate corporations have to date evolved very differently; companies in the defense and aerospace business have shown successful results.

GREAT BRITAIN

Great Britain has focused its national space policy on two areas, satellite telecommunications and remote sensing.

---

*Telespazio operates the space center in Fucino under contract to the government. The center serves as the point of contact for NASA and NOAA for the Earthnet program; it is also the national sales agent for Landsat and SPOT data.

43

Institutions

In November 1985 the activities of the various ministries and institutions were combined under a single national space agency, the British National Space Centre (BNSC). At its headquarters in London, it has about 40 employees from the relevant ministries and from industry; the technological research centers (in Farnborough and Chilton) employ an additional 240 people [57]. The declared goal of BNSC is to develop a broad and long-term national space program that includes individual projects and participation in such international programs as the ESA Columbus project.

Britain is receptive to bilateral cooperative activities, including countries outside ESA. An important country in such cooperation is the United States.

Activities

In the remote sensing segment, the major activities are

- Development of an active microwave instrument to deliver radar images of the ocean surface and provide wind and wave information. It will be used on the ESA satellite ERS-1.

- Participation (together with the United States) in the Canadian Radarsat program for all-weather imaging of land, ice, and ocean data.

The main thrust of space activities in Britain will remain satellite communications technology; approximately 47 percent of the funding has been in this area. However, earth observation and free-flying platforms are receiving increasing importance [58]. Whether other fields, such as microgravity utilization and in-orbit infrastructure development, will assume greater significance and will be pursued more strongly in the future depends largely on the national space program, which is still to be defined. The existence of the HOTOL project, an aerospace plane being studied by industry, has increased activities in propulsion and launcher development, although the British government has not funded HOTOL beyond October 1988 [59].

Research

Other than the above activities, no significant civilian remote sensing research programs are known (there are several military research programs). The National Remote Sensing Centre (NRSC) in Farnborough is responsible for research, application, and sale of remote sensing data. NRSC also supports equipment makers by marketing their products.

Budget

In both 1985 and 1986, the government spent nearly £100 million for space science and civil space R&D activities, over 75 percent of which was contributed to ESA [47]. The budget for 1987 increased to £117 million, of which £91.5 million (78 percent) was contributed to ESA [47]. For space telecommunications projects, industrial sales are nearly four times government expenditures (nearly £200 million in 1986), suggesting that the government priorities are correct [58].

Private Sector

Two private British companies, Hunting Surveys and Consultants and Nigel Press Associates, have recognized the value of satellite remote sensing data and are utilizing the data commercially.

Hunting Technical Services, a division of Hunting Surveys and Consultants, has considerable experience in land use management in the developing countries [59]. It works closely with the NRSC to develop the British market for regional and community authorities. In addition to satellite data, another division of the company, Hunting Surveys, offers aerial photographs with a ground resolution of 2 meters.

Hunting Technical Services processes about 100 Landsat tapes per year. Because of its own large data processing capacity, the company has been interested in acquiring raw data. Since U.S. companies have attempted to dominate the European market by controlling the European Earthnet system, Hunting Technical Services, together with other national points of contact (DFVLR in Germany, Satimage in Sweden, SPOT Image in France, and Telespazio in Italy), has formed Eurimage, the European consortium in 1985 responsible for the commercialization of Earthnet activities being considered by ESA. For private companies, the marketing concept of SPOT Image is considerably more attractive than the EOSAT concept for Landsat utilization [6].

Nigel Press Associates was founded to explore and prospect for oil [60]. The company analyzes satellite data to detect and locate promising oil formations. For this purpose, it has developed its own digital data processing system. Nigel Press Associates competes with the NRSC as a British point of contact for marketing and sale of SPOT data. The company perceives government efforts to develop the market for satellite data to be very obstructive, even though the NRSC is responsible only for sale of SPOT data sensed over British territory and Nigel Press Associates markets and sells all other SPOT data [59].

Although the NRSC has excellent technical facilities and very competent staff, it is still considered important to separate governmental from private industry activities: the government maintains data archival (NRSC houses 21,000 tapes), and private companies provide technical consultation [47,58].

Nigel Press Associates founded the British Association of Remote Sensing Companies (BARSC) in February 1986 to develop new markets for remote sensing. With the exception of a few public corporations, BARSC is mainly comprised of 16 private consulting firms.

No national program currently exists for commercialization in remote sensing applications. However, training programs supported by the government are being offered at the National Archive of Data in the NRSC in Farnborough. These programs are designed to demonstrate the various possibilities for commercial utilization of remote sensing data. Here the possibilities of commercial application of SPOT data are considered more limited even though the data is in many ways superior to Landsat data [58].

Comments

If commercialization of satellite remote sensing in Britain is to succeed based on enhancement and sales of the data, it will be mainly because of the flexibility and dynamism of the private sector. Governmental framework conditions are fairly limited and the budget for space research is generally small. Education, training, and consulting at the NRSC apparently have been used mainly by federal planning and cartography agencies, the oil companies, and value-added processors. Whether private sector dynamism can be injected and sustained in such a lean situation certainly cannot possibly be assessed in general terms. Likewise it is difficult to draw any conclusions about the effect of aggressive marketing of remote sensing data by a private company on the overall success of the government's hands-off approach to commercialization. Certainly, Britain's trade and information ties with many African and Asian countries provide some advantages in worldwide marketing of remote sensing data. But in the long term, these countries cannot be a substitute for the domestic market, since the operators of both currently available satellite systems (Landsat and SPOT) are increasingly locating their sales and marketing centers in the developing countries.

EUROPEAN SPACE AGENCY

As previously mentioned, the European countries are carrying out a substantial number of their space activities within projects administered by the European Space Agency (ESA). The following section discusses the infrastructure created by ESA as distinct from the national infrastructure of the individual ESA members.

Goals

In Europe about 13 years ago, ESA was formed as a multinational space agency by combining two previous organizations, the European Space Research Organization (ESRO) and the European Launcher Development Organization (ELDO) [1,61]. Its goal is to integrate the individual European space programs [62]. Originally, it was also intended for multinational management of large projects; however, labor is generally divided such that a single country (or its industry) under ESA management takes the lead in a specific project (for example, France for development of Ariane launch vehicle, Germany for Spacelab and Columbus, Great Britain for communications satellite systems) [63]. As a rule, leadership of a project accrues to the country that provides the largest financial contribution to the project (France, 62.5 percent for Ariane; Great Britain, 56 percent for the satellite systems; Germany, 56 percent for Spacelab and 38 percent for Columbus).

The ESA programs also serve to improve the worldwide competitiveness of European industry. The projects are supported by the member states in two ways: through mandatory contributions and through participation in optional projects such as Columbus, EURECA, and DRS. Although the size of the mandatory contribution of the member country is based directly on its GNP, each member can choose whether to participate, and at what rate, in the optional projects suggested by the ESA Council.

Institutions

For launch systems, ESA has financed the Kourou spaceport in French Guyana, which currently has two operational launch pads for Ariane launches, ELA-1 and ELA-2. In November 1988, construction was started on a third launch complex, ELA-3, to meet the requirements of Ariane-5. It is scheduled to be operational in 1993 [63]. This will allow ESA about ten annual Ariane launches with the minimum interval of one month between successive launches [64].

ESA and NOAA have agreed on the receipt and distribution of Landsat and meteorological data within the Earthnet Program of ESA. The Earthnet Center is located in the European Space Research Institute (ESRIN) in Frascati, Italy (DFD is the corresponding national point of contact in Germany.) The data are distributed not only to ESA members, but also to third parties [47].

Activities

One of the important ESA remote sensing programs is the Earthnet Program. This project is concerned with receiving and processing Landsat data, developing ERS-1 and ERS-2, and participating in joint experimental missions with other partners such as earth observation from Spacelab. Regarding Landsat data, ESA has signed an MOU with NOAA to obtain the data through the ground station in Fucino near Rome into the European Earthnet System [65]. A special office has been established within ESA to develop the market for commercial utilization of the data.

ESA is also developing research satellites and sensors for remote sensing. The first of these is ERS-1 (European Remote Sensing Satellite), which is being built jointly by European and Canadian companies [66]. ERS-1 is planned for a 1990 launch and it will be equipped with an active microwave instrument (6.5 GHz), active radar altimeter (13.7 GHz), laser reflector, infrared orbit sensor, microwave sensor, and an instrument for the Precise Range and Range Rate Experiment (PRARE) [67]. The data obtained from these instruments can be used in various ways:

- Exact determination of wind areas

- Precise images of ocean and sea waves

- Radar imaging of the Earth's surface (resolution of 30 meters)

- Sea surface temperature measurements.

This data can be made available to the user in three different forms:

- Essentially unenhanced raw data (with some additional helpful information)

- Preprocessed data with wind and wave information and low-resolution radar images (approximately three hours after receipt)

- Fully processed and corrected data, if time is available.

A final decision has not yet been reached on the marketing of ERS-1 data. ERS-1 is not seen as competition for SPOT or Landsat, rather it will complement SPOT and Landsat data.

The planned marketing activities for ERS-1 data could be decisively set back by the demand of the ESA members for free access to the data. Hence, future ESA space projects will be planned from idea to commercialization to satisfy both research and user needs [68].

Budget

ESA currently has a total of about 1400 employees. Table 8 gives proposed ESA spending plans through 2000. Individual member contributions to the mandatory activities for 1987 are given in Table 9. From 1988 to 1990, the contribution scales for the mandatory activities were changed for full members; these are shown in Table 10 [69,70]. These spending plans are summarized in Figure 4, which clearly shows that Columbus (22.5 percent) and STS (32.6 percent) activities receive the bulk of ESA funding, although earth observation also receives due emphasis with about 10 percent of total funding.

Individual member contributions to the optional remote sensing programs are also interesting [48]. As Table 11 shows, for the two Sirio-2 projects, Italy is the main contributor at 72 percent (62 percent for the exploitation phase). Meteosat has nearly 22 percent French financing, followed by Germany with a 21 percent share. Germany leads in contributions to all other projects with over 25 percent share of the respective total budgets (ERS-1, Phase B—Extension at 35 percent).

TABLE 8. *Profile of Expenditures for All ESA Programs from 1987 to 2000*[a]

Programs	1987	1988	1989	1990	1991	1992	1993	1994	1995	1996	1997	1998	1999	2000	Total
MANDATORY PROGRAMS															
General Budget	124.7	120.0	126.0	138.1	148.0	161.0	167.0	170.0	170.0	170.0	170.0	170.0	170.0	170.0	2174.8
Associated with General Budget	66.1	55.0	52.6	57.7	62.9	67.1	67.1	68.1	68.4	69.7	70.8	71.8	72.9	74.0	924.2
Science	170.3	176.8	185.3	196.7	206.4	216.7	227.9	232.0	232.0	232.0	232.0	232.0	232.0	232.0	3004.1
TOTAL	361.1	351.8	363.9	392.5	417.3	444.8	462.0	470.1	470.4	471.7	472.8	473.8	474.9	476.0	6103.1
USER PROGRAMS															
Science	170.3	176.8	185.3	196.7	206.4	216.7	227.9	232.0	232.0	232.0	232.0	232.0	232.0	232.0	3004.1
Technology Programs	6.7	4.4	12.8	21.0	23.9	27.3	30.3	32.4	32.4	32.4	32.4	32.4	32.4	32.4	353.2
Earth Observation	174.1	197.9	228.3	224.2	234.9	239.7	237.2	236.9	230.7	252.3	255.9	251.4	261.1	267.4	3292.0
Microgravity	13.0	46.9	78.1	110.0	110.0	110.0	110.0	120.0	130.0	142.0	142.0	142.0	142.0	142.0	1538.0
Telecommunications (excluding DRS)	289.6	248.5	210.5	185.6	192.7	179.1	153.7	158.8	172.0	204.0	224.0	213.7	162.9	162.9	2758.0
TOTAL	653.7	674.5	715.0	737.5	767.9	772.8	759.1	780.1	797.1	862.7	886.3	871.5	830.4	836.7	10945.3
SSP + STS															
Approved Space Station and Platform (SSP)	195.3	63.7	38.9	24.4	15.0	3.0									340.3
Approved Space Transportation Systems (STS)	446.4	363.0	69.6												879.0
TOTAL	641.7	426.7	108.5	24.4	15.0	3.0									1219.3

TABLE 8. *Continued*

Programs	1987	1988	1989	1990	1991	1992	1993	1994	1995	1996	1997	1998	1999	2000	Total
SPACE INFRASTRUCTURE															
DRS			5.0	7.0	36.0	65.0	82.0	109.0	121.0	115.0	110.0	52.0			702.0
Columbus		158.0	258.0	324.0	418.0	427.0	453.0	443.0	423.0	350.0	140.0	62.0			3456.0
Ariane-5		72.0	410.0	485.0	510.0	565.0	565.0	430.0	350.0	109.0					3496.0
Hermes		76.4	170.0	318.0	322.0	380.0	464.0	535.0	565.0	560.0	525.0	475.0	39.0		4429.4
TOTAL		306.4	843.0	1134.0	1286.0	1437.0	1564.0	1517.0	1459.0	1134.0	775.0	589.0	39.0		12083.4
OPERATIONS															
DRS Operations											5.0	10.0	10.0	10.0	35.0
Columbus Launches							10.0	30.0	60.0	100.0	73.0	44.0			317.0
Columbus Operations							20.0	30.0	42.0	130.0	365.0	395.0	520.0	504.0	2006.0
Columbus Utilization Preparation Program		3.4	11.6	10.0	30.0	40.0	20.0	10.0	10.0						135.0
Hermes Demonstration										31.0	42.0	43.0	116.0	38.0	270.0
Hermes Operations												24.0	84.0	104.0	212.0
TOTAL		3.4	11.6	10.0	30.0	40.0	50.0	70.0	112.0	261.0	485.0	516.0	730.0	656.0	2975.0
FUTURE PROGRAMS															
DRS-2													50.0	75.0	125.0
EMSI-II		3.0	4.0	5.0	6.0	8.0	10.0	12.0	20.0	20.0					88.0
EMSI											100.0	180.0	370.0	373.0	1023.0
Ariane-4 PAPA			12.0	20.0	25.0	25.0	25.0	25.0	12.0						144.0
Ariane-5 PAPA										30.0	50.0	50.0	60.0	60.0	250.0
FESTIP		5.0	8.0	10.0	10.0	10.0	20.0	20.0	25.0						108.0
Future Launchers										80.0	90.0	180.0	300.0	373.0	1023.0
TOTAL		8.0	24.0	35.0	41.0	43.0	55.0	57.0	57.0	130.0	240.0	410.0	780.0	881.0	2761.0

[a] MAU in 1986 currency units.

Legend:
DRS = Data Relay Satellite
EMSI = European Manned Space Infractructure
FESTIP = Future European Space Transportation Investigation Program

49

TABLE 9. *Mandatory Contributions from Member States to ESA's Budget for 1987*

Member state	Percent share	Budget share in MAU[a]
Austria	0.9	9.3
Belgium	4.0	39.4
Denmark	1.2	11.8
France	25.9	257.2
Germany	23.8	235.7
Ireland	0.2	2.2
Italy	15.4	152.5
Netherlands	3.9	38.8
Norway	0.8	8.3
Spain	3.2	31.7
Sweden	2.8	27.8
Switzerland	2.0	19.7
United Kingdom	13.3	132.2
Canada (associate member)	2.3	24.9
Finland (associate member)	0.2	1.7

[a] MAU = million accounting units.

TABLE 10. *Contribution Scales for Mandatory ESA Activities for 1988 through 1990 (percent)*

Member state	1988	1989	1990
Austria	2.34	2.33	2.31
Belgium	3.09	3.07	3.05
Denmark	1.97	1.95	1.94
France	18.37	18.27	18.15
Germany	23.54	23.41	23.25
Ireland	0.40	0.40	0.50
Italy	14.07	14.00	13.91
Netherlands	4.96	4.94	4.91
Norway	2.00	1.98	1.97
Spain	5.17	5.69	6.22
Sweden	3.55	3.53	3.50
Switzerland	4.03	4.01	3.98
United Kingdom	16.51	16.42	16.31
TOTAL	100.00	100.00	100.00

Private Sector

Landsat data is currently sold in Europe through the Earthnet system of ESA. Like NASA, ESA is planning to transfer the marketing of at least some Landsat systems to the private sector.

In March 1980, a European "private corporation" called Arianespace was established on French (CNES) initiative to produce, finance, market, and launch the Ariane vehicles developed under French leadership [71]. Arianespace will operate as a commercial profit-oriented company responding rapidly to the demand for launch services. This will also ensure the competitiveness of Ariane with the Space Shuttle and American ELVs. Table 12 lists the shareholders of Arianespace.

Arianespace has an agreement with the Centre Spatial Guyanais (CSG) in Kourou, French Guyana, that defines the functions of the CSG for launches of space vehicles provided by Arianespace. The launches are completely paid for by Arianespace; the CSG only provides the usual ground operations facilities and services.

TABLE 11. *Contributions of the Countries Considered to Key ESA Optional Remote Sensing Programs (percent)*

Program	1	2	3	4	5	6	7	8
Germany	9.00	11.15	24.00	34.87	26.60	17.57	20.00	21.00
France	7.50	16.49	18.31	16.55	21.58	16.53	18.17	22.00
Italy	72.39	62.04	10.61	12.20	11.32	11.32	11.00	11.00
United Kingdom	1.83	2.27	13.34	18.17	13.88	16.83	15.18	14.40

Legend: 1 = SIRIO-2   5 = ERS-1, Phase C/D
2 = SIRIO-2, Exploitation Phase   6 = ERS-1, Phase E
3 = ERS-1, Phase B   7 = Earth Observation Preparatory Program
4 = ERS-1, Phase B Extension   8 = METEOSAT

The Ariane series of vehicles (Ariane-1 through 4) are capable of launching all current and planned satellites. The newest vehicle, Ariane-4, successfully launched on June 15, 1988, has six configurations and can lift 6,000 kg to LEO and between 1,900 and 4,200 kg to GTO [72]. Ariane-5 has been approved for funding by ESA and will be available for commercial use in 1996. It will have the capacity to place 6,500 kg into GTO in dual or triple launch configuration [72]. Arianespace offers its launch services on a commercial basis to all customers with payloads for peaceful uses of space; however, French, British, and NATO military

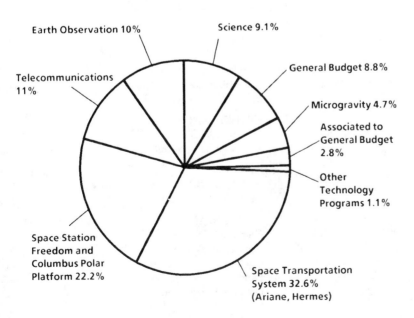

FIGURE 4.
Profile of Expenditures in the ESA Long-Term Plan for All Programs from 1987 to 2000 (MAU in 1986 e.c.)

satellites are also accepted for launch. Political control over the launches is exercised by the ESA Executive Council through a relevant agreement between ESA and Arianespace.

Arianespace captured about half the Western market for satellite launches in 1985 and as of June 1987 had outstanding orders to launch 46 satellites through 1991. The annual turnover is expected to be FFr 2.5 to 3 billion [71].

ESA is presently studying the feasibility of commercial utilization of ERS-1 [67]. ERS-1 is not seen as competition for SPOT and Landsat, but rather as a supplemental information source. Potential users are shipping companies and oil prospectors, primarily because of its radar instrumentation.

TABLE 12. *Percent Ownership of Shareholders of Arianespace*

GERMANY	**19.6**	SFIM	0.1
Dornier System GmbH	2.8	SODETEG	0.1
ERNO Raumfahrttechnik GmbH	5.2	B.N.P.	0.01
Maschinenfabrik Augsburg	7.9	Société Financière Auxiliaire	0.49
Nürnberg (MAN)		Crédit Lyonnais	0.5
Messerschmitt Boelkow Blöhm (MBB)	2.8	OPFI PARIBAS	0.4
Dresdner Bank	0.3	VALORIND	0.4
Bayerische Vereinsbank	0.3	Banque Vernes et Commerciale	0.02
Westdeutsche Landesbank	0.3	de Paris	
		Diverses Personnes Physiques	—
BELGIUM	**4.4**		
Études Techniques et	1.3	**UNITED KINGDOM**	**2.4**
Constructions Aerospatiales		Avica Equipment Limited	0.3
(ETCA)		British Aerospace	0.95
Fabrique Nationale Herstal	0.7	Ferranti Limited	0.95
Société Anonyme Belge de	2.4	Midland Bank Limited	0.2
Constructions Aéronautiques			
(SABCA)		**IRELAND**	**0.25**
		ADTEC Teoranta	0.15
DENMARK	**0.7**	Aer Lingus	0.1
Christian Rovsing AS	0.5		
Copenhagen Handelsbank	0.2	**ITALY**	**3.6**
		Aeritalia	1.1
SPAIN	**2.5**	Selenia Spazio	0.8
Construcciones Aeronauticas	1.9	SNIA BPD SpA	1.4
SA (CASA)		SI.EL.SpA-A	0.3
Sener	0.6		
		NETHERLANDS	**2.2**
FRANCE	**59.25**	Fokker VFW B.V.	1.9
Aérospatiale	8.5	Algemene Bank Nederland N.V.	0.3
L'Air Liquide	1.85		
Centre National d'Études	34.0	**SWEDEN**	**2.4**
Spatiales (CNES)		Saab-Space AB	0.8
Comsip-Enterprise	0.1	Volvo Flygmotor Aktiebolag	1.6
Crouzet S.A.	0.1		
Compagnie Deutsch	0.1	**SWITZERLAND**	**2.7**
Intertechnique	0.1	Compagnie Industrielle	0.15
Matra	3.6	Radioèlectric (CIR)	
SAFT	0.1	Contraves A.G.	2.15
Société Européenne de	8.5	Fabrique Fédéral d'Avions	0.1
Propulsion (SEP)		Union de Banques Suisses	0.3
SFENA	0.1		

For satellite remote sensing, Europe has an outstanding scientific base, adequate governmental R&D financing, a mature aerospace industry, coordination among the ESA member states, a positive long-term economic outlook for commercialization, and the political framework for limited European autonomy in launch systems. Infrastructure is not lacking; the situation could be improved only by internal markets with lucrative short-term payoffs.

Comments

The main goal of U.S. space policy is to retain its leadership in space technology and its application for peaceful purposes in and out of the atmosphere [19].

UNITED STATES OF AMERICA

Remote sensing institutions in the United States are described under the following general organizations [2,73,74]: federal government organizations, state and local government organizations, private industry, and universities, academic, and other nonprofit organizations.

Institutions

Federal Government Organizations.   Remote sensing activities conducted at the federal government level are discussed under two types of organizations, R&D organizations and satellite system operation organizations.

The major R&D organization in remote sensing is NASA. NASA plans, budgets, and coordinates all research and development in remote sensing, including Metsat, Landsat, and ocean and ice satellite systems. NASA field centers conduct major parts of the research. The major NASA centers involved in remote sensing are

- Goddard Space Flight Center, Greenbelt, Maryland

- Jet Propulsion Laboratory, Pasadena, California (operated by the California Institute of Technology under contract to NASA)

- National Space Technology Laboratory, St. Louis, Missouri

- Ames Research Center, Moffet Field, California.

NASA headquarters also supports research activities through many universities, nonprofit organizations, and various small and large companies.

NOAA has the primary responsibility for the operation of remote sensing satellite systems, including the Metsat and the Landsat systems. The organization within NOAA with satellite responsibility is the National Environmental Satellite, Data, and Information Services (NESDIS), located in Suitland, Maryland.

In addition, several components of NOAA are involved in R&D and use satellite data: National Weather Service, National Ocean Service, National Marine Fisheries Service, and Office of Oceanic and Atmospheric Research.

State and Local Governments.   Many state and local governments use remote sensing in various activities. Some 22 state agencies have instituted resource information programs in broad-based GIS [9,75]. Associated with state and local organizations are usually many small companies that get started with a specific remote sensing specialty. These companies support the state remote sensing activities through contracts. Most of the state and local governments have been encouraging small business and high technology companies to locate within their boundaries.

Private Industry.   A significant amount of U.S. private industry activity in remote sensing is conducted through the Geosat Committee, Inc., in San Francisco, California. The Committee was formed in 1976 and is sponsored by almost 100 oil, gas, mineral, and engineering companies. The goal of the committee is to present to NASA and NOAA the considered geological recommendations of the nation's largest single group of users of satellite imaging data. It is a lobbying organization and coordinates and/or participates in specialized remote sensing workshops and symposia. Special studies are conducted by staff of the participating members.

Universities, Academic, and Other Nonprofit Organizations.   Forty-five such organizations are active in practically all areas of remote sensing [75], principally research and development. Most of their funding comes from government organizations, some from industrial organizations.

The Institute for Technology Development's (ITD) Space Remote Sensing Center in Hancock, Mississippi, is one of 16 Centers for the Commercial Development of Space (CCDS) established by NASA [74], and one of only two in remote sensing. The budget of the Center for the first year (1985–86) was about $2.5 million in cash and in-kind services. The Center's objective is to serve as the national focal point for activities designed to promote the effective commercialization of remote sensing technology; projects have included agriculture, forestry, and land management. The Center plans research in sensor technology, analytical data interpretation techniques, and data handling capabilities in three sectors of the remote sensing industry: data supplies, information users and suppliers, and software and hardware systems suppliers. The Center's membership includes industry, government, and other universities.

The Center for Mapping at the Ohio State University in Columbus, Ohio, one of the four CCDSs selected by NASA in the second round in 1986, is the other CCDS in remote sensing. Its objective is to develop and integrate the following four different capabilities into commercial opportunities for space remote sensing systems:

- Timely collection of the necessary input data, especially remotely sensed data that accurately locates surface features in terms of their geographic coordinates

- Merging these input data collected by various means and from different sources

- Converting the data into useful information through the use of application and process models

- Providing the derived information in a timely manner to individuals and organizations in most need of the information.

The focus on real-time satellite mapping is in the following areas: market needs in land and water resources, farm management, energy and power production, offshore environment monitoring, and disaster monitoring; development and implementation of digital mapping and land information systems; and study of what equipment should fly and on what vehicles.

The Center currently has nine partners with an estimated total five-year budget (1986–1991) of $12 million; about $5 million is from NASA, $3.7 million from the state or university, and the remainder from industry.

Several federal government agencies use satellite remote sensing data and services. These agencies also conduct their own research to determine the type of sensors or satellites they may need in the future [2,76]. These agencies include the Department of Agriculture (USDA), Agency for International Development (AID), Department of Transportation (DOT), Army Corps of Engineers (ACE), Department of Defense (DOD), and Department of the Interior (DOI).

The USDA is responsible for monitoring the production of food and fiber in the United States. Their major requirements include weather data, information, and services. Specific data and information requirements include precipitation, temperature, snow, wind, soil moisture and temperature, cloud cover, solar radiation, radiation and pressure.

AID's mission is to foster the growth of developing countries by providing a wide range of assistance programs aimed at enhancing basic capabilities in such areas as food security, infrastructure and institutional development, and environmental protection. Their major remote sensing requirements include

- Early warning of natural disasters such as crop failure, drought, and flooding

- Agricultural, climatic, and environmental monitoring to help in planning and management of resources

- Meteorological and weather modification programs, and disaster management.

The types of applicable data include surface wind data, cloud motion and cover, sea surface temperature, precipitation, surface energy fluxes, and historical satellite data.

The DOT has two operational components with major interest in remote sensing: the Coast Guard, responsible for offshore missions in search and rescue in the United States, and the Federal Aviation Administration (FAA), responsible for ensuring the safety of U.S. aviation activities. The major Coast Guard remote sensing requirements are for oceanographic, sea ice, and iceberg data. The FAA requires information on weather and severe storms, clouds and visibility, lightning, satellite imagery, surface temperature, and information for forecasting and for various aviation charts.

The Army Corps of Engineers, through its Civil Works Program, carries out a nationwide water resources planning, construction, and operations effort in close cooperation with a wide range of public and private organizations. Specific project areas include navigation improvements related to the nation's rivers and harbors, flood damage reduction, hydropower generation, beach erosion control (coastal and lakes), fish and wildlife management, and environmental enhancement. Applicable remote sensing data and services include:

- Soil moisture information

- Meteorological and weather data

- Flood location

- Wind, ice, and snow mapping, including precipitation information

- Land cover maps

Activities

- Wildlife habitat identification

- Geologic structure mapping

- Predictions of water quality.

The DOD mission requires global capabilities. Satellite remote sensing is of great interest. DOD conducts its own R&D, develops and operates the system either directly or with the help of contractors, and is a major user of remote sensing data. DOD activities will not be discussed here in any detail because DOD is not active in commercialization. However, portions of DOD satellites may apply since they provide data for civilian use; in particular the microwave imager planned on the Defense Meteorological Satellite Program (DMSP) will provide ice imagery, and N-ROSS will provide oceanographic data. These imagery and data will become available through NOAA for civilian distribution in the future.

Through its various bureaus and offices, DOI has direct administrative responsibilities for about 40 percent of the total land and continental shelf areas of the United States. DOI is also responsible for mapping and resource appraisal for the entire country. DOI uses satellite remote sensing data in various applications such as:

- Reclamation of surface mines

- Monitoring the effects of acid rain

- Detection of thermal pollutants

- Detection and monitoring of oil pollution

- Regional and general geologic mapping

- Detailed delineation of geographic structures

- Determination of rock compositions

- Geobotanical studies

- Monitoring mountain snowpack

- Monitoring lakes and ponds

- Determination of shallow aquifers

- Agricultural crop classifications

- Vegetation mapping and changes.

Spatial resolution requirements for these activities are in the range of 10 to 80 meters.

Research   Research and development of sensors and scanners is shared by NOAA and NASA. The following instruments have been or are being developed [2]:

- Advanced very high resolution radiometer (AVHRR) for cloud, hydrologic, and oceanographic parameters

- High resolution infrared radiation sounder (HIRS) to obtain vertical temperature profiles from the Earth's surface

- Visible infrared spin scan radiometer (VISSR) to track the hourly motion of typhoons, hurricanes, snow, blizzards, and heavy rain storms

- Large format camera (LFC) to provide precise stereoscopic photographic imagery of Earth at very high resolution (similar to the German metric camera)

- Coastal zone color scanner (CZCS) to obtain ocean surface information

- Solar occultation absorption spectroscopy sensor (SOASS) with a high-resolution scanning spectrometer to measure atmospheric spectra

- Differential absorption LIDAR, Raman scattering LIDAR, and Doppler-shifted LIDAR

- Airborne visible/infrared imaging spectrometer (AVIRIS)

- Multilinear array instrument (MLA), which is similar to but more capable than the SPOT sensor

- Electrically scanning microwave radiometer (ESMR) to obtain information under nearly all weather conditions on sea surface temperature, ice cover, and other hydrologic features

- SMMR, SSM/I, LFMR, which are also microwave radiometers

- Active cavity radiometer (ACR) to measure solar radiation

- Microwave altimeters to map ocean topography

- Scatterometers to accurately measure ocean driving force (winds) and topography changes

- SAR with the Shuttle Imaging Radar (SIR) program to conduct geoscientific investigations (SIR-D will be flown on the Columbus Polar Platform).

Within the NASA remote sensing centers and the CCDSs, there are plans for sensor development in addition to other applied R&D activities. Independent R&D activities are also pursued by the U.S. Air Force and the U.S. Navy.

Budget

The NASA R&D budget for fiscal year 1989 is $4.266 billion, which includes $413.7 million for earth sciences [77]. Within this broad area are TOPEX ($83 million), scatterometer ($10.6 million), earth science payload instrument development ($46.4 million), ocean processes research and analysis ($20.8 million), and land processes research and analysis ($19.9 million); the remaining funding is for atmospheric research and analysis, as well as for such noncommercial research topics as the geodynamics program and climate and radiation research. For fiscal year 1990, the budget request is $434.3 million for earth sciences, representing a 5 percent increase over 1989 [77]. Despite the need for significant funds in the R&D budget for other activities such as planetary exploration, NASA's budget for earth sciences has increased significantly since 1986 ($365 million). However, to support the major activities in the EOS core program through the 1990s, NASA will have to maintain a funding level of $250 million annually [78].

More important for commercialization are the expenditures for the CCDSs:

- The total budget for the ITD Center was approximately $2.5 million in the first year (1985–1986). NASA provides about 50 percent of the support.

- The Center for Mapping at Ohio State University currently has nine members. Its planned five-year budget (1986–1991) is $12 million, of which NASA will provide about $5 million.

Also of interest is the budget of NOAA, which had managed the Landsat program until 1985 when its transfer to the private sector was announced. Under pressure to reduce overall spending, funding requests for 1987 for satellite remote sensing programs were lower than in 1986 ($263.5 million versus $331 million). These reductions will affect various future satellite systems for atmospheric and climate research. As noted below, funding for EOSAT has recently been approved for continued operation of the Landsat system through the 1990s and beyond.

Private Sector   Many companies are involved in remote sensing activities. The largest of these are the aerospace companies General Electric, Lockheed, Martin Marietta, Boeing, TRW, RCA, Hughes, and Rockwell. Although some of them market a specific remote sensing capability, the majority are interested in developing and building spacecraft and ground processing facilities. These will not be discussed any further.

Another group of companies are the user companies, such as those with oil, gas, pipelines, and marine transportation interest and activities. In many cases, these user companies depend on other companies (either vendors of hardware and software or value adders) to provide them with remote sensing products tailored to their needs. Exceptions include many of the major oil companies that have set up their own remote sensing laboratories; they buy remote sensing raw materials and equipment and develop their own software or value-added products. Examples include Exxon, Mobil, Texaco, Arco, and Shell. The smaller oil companies and other users are still dependent on the value-added companies to serve their needs.

More than 125 U.S. companies provide value-added services using Landsat and SPOT data [8]. Also, over 50 companies produce information processing hardware and software for remote sensing applications [9]. These figures represent a significant increase in the number of U.S. companies involved, since a 1980 report [75] identified about 60 companies involved in Landsat activities.

The Landsat system has been commercialized. In May 1985, the Department of Commerce (DOC) selected Earth Observing Satellite Company (EOSAT) as the winner according to the terms of the Landsat Commercialization Act of 1984. EOSAT is a corporation formed by RCA and Hughes Aircraft Corporation specifically to market Landsat data and construct, own, and operate follow-on Landsat satellites. EOSAT, DOC, and the Office of Management and Budget (OMB) agreed in 1985 over the amount of subsidy and financial risk to EOSAT. This subsidy consisted of $250 million plus launch costs (a total of $290 million). EOSAT agreed to build and launch two satellites whether or not the market has developed to support a profit-making business. For FY 1987, Congress had approved $72 million for Landsat [79]. More recently, DOC and EOSAT have signed a contract to continue the Landsat satellite program into the 1990s. The $220 million agreement provides for the commercial development and construction of Landsat-6 and related ground systems. In addition, EOSAT will continue worldwide market development for Landsat products and the operation of Landsats-4 and 5 [8].

Landsat-6 development activities will begin immediately, with EOSAT receiving an initial funding allocation of $62.5 million of the total contract award for the program restart.

The new contract calls for a net cost to the U.S. government of $209.2 million, as agreed to by the Administration and Congress in October 1987. The contract specifies that

- Landsat-6 will be an advanced TIROS-N design for launch aboard a Titan-2 ELV from Vandenberg Air Force Base, California, in June 1991.

- The government will provide a total of $220 million for the construction of Landsat-6 and associated ground systems.

- EOSAT shall return to the government the first $2.5 million of marketing revenues each year, beginning in 1988, until $10.8 million has been returned.

Under the new agreement, Landsat-6 will carry an enhanced thematic mapper (ETM) sensor to provide imaging capability in seven spectral bands with 30-meter ground resolution and a 15-meter panchromatic capability. In addition, EOSAT and NASA have developed a sea-wide field sensor (Sea-WiFS) to provide wide area, low-resolution (1 km) ocean color and temperature information. EOSAT is investigating the addition of a 5-meter resolution, full-color satellite tracking and reporting (STAR) sensor for the Landsat-6 mission.

The 1984 Act also gave NASA the continued responsibility for research and development of remote sensing sensors, systems, and techniques. It encourages NASA to cooperate with other federal agencies and with public and private research entities both nationally and internationally in the pursuit of R&D in remote sensing. The Act further encourages other federal agencies to conduct their own research programs in remote sensing.

Comments

Utilization of remote sensing data has the longest tradition and the largest market in the United States. Despite the recent uncertainties regarding the operation of the Landsat system, it is now clear that Congress is determined to operate a U.S.-owned land remote sensing system in the United States. The first stage of the planned commercialization (transfer to the private sector), however, has unexpectedly created major problems. This suggests that the cost-benefit ratio between the federal agency and the private operator was estimated incorrectly. In an effort to commercialize the system all at once, EOSAT was saddled with an enormous initial risk without any assurance of additional federal support in the event of unexpected difficulties. Meanwhile, the U.S. government has approved additional funding for EOSAT. The commitment of NASA exclusively to sensor development very clearly illustrates the difficulties in reaching a satisfactory agreement between the interested parties.

SPOT Image offers customers an alternate data source; it will be able to penetrate and maintain a share of the U.S. market. The disadvantage, however, of the SPOT satellites is their fewer spectral bands compared with Landsat.

JAPAN

Japan plans to launch its own satellites as well as using foreign satellites for remote sensing. For the future, the Japanese space program is aimed at developing Japanese space technology as equally competitive with U.S. and European space technology [80].

Institutions and Activities

Following the successful launch of the H-1 vehicle on August 13, 1986, Japan has accelerated the development of the H-11 vehicle with the first test flight planned for 1992. H-11 will have the capacity to place 8,000 kg into LEO and approximately 3,860 kg into GTO from the Tanegashima Space Center launch complex [81]. The goal is to reduce the launch costs to reasonable and competitive levels, as well as to provide its own flight opportunities (for microgravity experiments, as well as remote sensing, independent of Europe or the United States). In the next ten years, Japanese competitiveness in space utilization for remote sensing is expected to increase significantly.

The Japanese National Aeronautics and Space Development Agency (NASDA) has developed the Marine Observation Satellite-1 (MOS-1), which will determine ocean conditions by sensing wave heights, ocean color, and surface temperature. MOS-1 was launched in February 1987 on an N-11 vehicle. The follow-on system MOS-1B, scheduled for launch in 1990, will have higher quality sensor capability [82]. NASDA is also increasing its efforts to develop uniform technology applicable to all earth observation satellites.

In cooperation with the Ministry of International Trade and Industry (MITI), NASDA is also involved in developing the Japan Earth Resources Satellite (ERS-1), which will be used to collect information on renewable and nonrenewable earth resources [83]. It is planned for launch in late 1990. Japan's goals for the ERS-1 mission are to test synthetic aperture radar (SAR) technology and its optical sensors. Further, information will be collected in geology, agriculture, forestry, fisheries, environmental pollution, and catastrophe prevention in keeping with the primary functions of the satellite to survey and locate Earth resources. In addition, coastal regions will also be monitored.

For Japan's remote sensing program, the following options will be actively implemented:

- Satellite data will be used by Japan and other foreign countries on a reciprocal basis.

- International joint projects for development of remote sensing technology will be supported.

- Cooperative ventures will be developed, initially with the ASEAN countries, to encourage the use of satellite data by the developing countries.

- Japan will participate in the program for earth observation from the Columbus Polar Platform, which is being emphasized in Europe and the United States.

In support of these activities, the next step of the earth observation program will be to validate technologies to be used on the Advanced Earth Observing Satellite (ADEOS) scheduled for launch in 1993 aboard the H-II rocket. ADEOS is a platform arrangement housing several sensors such as TIR, and it will be capable of changing its attitude and orbit.

Japan has developed the following remote sensing sensors for MOS-1:   **Research**

- Multispectral electronic self-scanning radiometer (MESSR) to measure sea and land cover
- Visible and thermal infrared radiometer (VTIR) to measure sea surface temperature and atmospheric vapor
- Microwave scanning radiometer (MSR) to measure atmospheric vapor over oceans, sea ice, and snow pack on land.

A synthetic aperture radar (SAR) instrument and an optical sensor (visible and near-infrared radiometer, VNIR) are presently being developed for ERS-1 in a joint project of the Science and Technology Agency (STA), which also includes NASDA and the MITI research group, the Agency for Industrial Science and Technology (AIST) [83].

In this section, the expenditures for remote sensing are compared with the total   **Budget**
expenditures for space and the total expenses for all R&D activities.
   The total budget for space activities was ¥141,780 million in 1988. This is a 15 percent increase over 1987 (¥123,200 million). Through the year 2000, the government is expected to invest about ¥6 trillion in space activities. The budget for remote sensing was ¥13,037 million in 1987. Thus, remote sensing accounts for about 11 percent of the total space R&D budget. Table 13 provides more detailed information on expenditures for remote sensing for 1987.

In 1970, Japan began a program, mainly through STA, to support the commercial   **Private Sector**
utilization of remote sensing data. This program was aimed at the commercial utilization of Landsat data. In July 1975, the Remote Sensing Technology Center

TABLE 13.   *Expenditures for Satellite Remote Sensing in Japan in Fiscal Year 1987 (million Yen)*

Agency and activity	Budget
NASDA:	
• Earth Observation Data Management	2,645
MITI:	
• Development of ERS-1	5,140
• R&D in remote sensing technology for exploration of nonrenewable resources	1,350
• Development of advanced sensor system for nonrenewable resources	400
Ministry of Transport:	
• Meteorological Agency (weather forecasting and operation of geostationary satellites)	3,245
• Maritime Safety Agency	255
Ministry of Construction:	
• Geographical Survey Institute (geodetic measurements)	2
TOTAL	13,037

of Japan (RESTEC) was established, and in September 1981, the Earth Resources Satellite Data Analysis Center (ERSDAC) was established to perform and promote remote sensing R&D. In October 1978, the Earth Observation Center (EOC) of NASDA was established for reception, processing, and distribution of Landsat data through RESTEC. SPOT data will be received through the ground station at Katsura. Private companies have increasingly, and now routinely, employed satellite data for general surveys, and then perform more detailed ground surveys. Since the domestic Japanese markets are too small, Japanese companies are making insignificant investments in the remote sensing data business in Japan.

The number of Japanese firms from the major industrial sectors interested in satellite remote sensing applications is shown in Figure 5. Not unexpectedly, firms in the service and energy industries are the most interested in using remote sensing. Less clear is the similarly large interest of general mechanical engineering companies. Presumably, the survey questionnaire, on which Figure 5 is based, did not distinguish clearly enough between hardware manufacture and utilization.

The position of the private sector with regard to satellite remote sensing can be summarized as follows. Since early 1970, Japanese competitiveness has improved significantly in remote sensing technology. The most urgent problem at present is to develop a market for utilization of remote sensing data.

FIGURE 5.

Distribution of Japanese Firms Interested in Satellite Remote Sensing by Industrial Sector

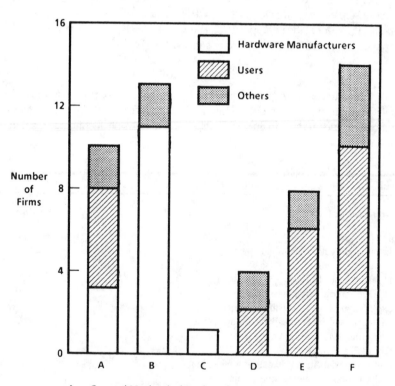

A   General Mechanical Engineering
B   Electrical and Electronic Industry
C   Chemical Industry
D   Mining, Iron and Steel, Metals Processing Industries
E   Energy Industry
F   Service Industry

The general interest of the private sector in Japan in space utilization has been aroused and is, in fact, growing rapidly in certain areas such as microgravity research. In the area of remote sensing, rapid and strong growth in utilization of satellite remote sensing data in Japan is *not* to be expected. Regional administrations, for example in the Tokyo, Osaka, and Hiroshima Prefectures, are now slowly beginning to use satellite data for various applications, such as observation of annual changes in land use, forestry, and monitoring environmental conditions. This market segment will increase only slightly and will be competing with aircraft remote sensing. The situation is attributed not so much to a lack of political and economic framework conditions, but to the geographical realities of Japan. Only 15.4 percent of the total land area of the Japanese islands is arable, the rest is mountains and wilderness. Hence, the domestic market for earth observation is essentially negligible. The situation is very different for ocean observation and location of raw materials. Here the data from both satellites MOS-1 and ERS-1 (planned launch in 1990) will find interesting applications worldwide as remote sensing knowledge increases. The instruments aboard MOS-1 are similar in some respects to the SPOT instruments, which will make it possible to compare results directly.

Comments

CHAPTER

7

Comparison of Existing Conditions in Specific Countries

In the previous chapters, the requirements and the framework conditions for the countries considered were described on a collective basis. This form of description provides an overall view of the system within which commercialization is partly included, but it does not permit comparison of the individual requirements or conditions. For this purpose, in this chapter the prerequisites and conditions for each country are directly compared based on the criteria developed earlier.

Availability of Launch Vehicles and Facilities. All the European countries considered have access to Ariane vehicles and the ESA launch facility in Kourou, both owned and operated by France, through ESA membership. Great Britain's launch facility at Woomera in Australia can be reactivated. Italy operates an offshore launch facility on San Marco Island, Kenya, for Scout class vehicles. Germany lacks indigenous facilities, and its ties to ESA have been ineffective in assisting private customers.

Japan will replace launchers based on U.S. technology in the early 1990s with the H-II, a vehicle developed entirely by Japan. Limited indigenous launch facilities are available at the Tanegashima Space Center.

The United States maintains a fleet of large-capacity ELVs (Titan, Delta, and Atlas), in addition to the Space Shuttle, which no longer accepts commercial satellite payloads. U.S. launch facilities are the most extensive in the world, more than adequate to meet current demand and provide for future growth.

As a result of the commercialization of the market for telecommunications satellites, all interested satellite users have access to all launch vehicles in the West, the Soviet Union, and China. Launch delays in the Shuttle program and in Ariane will take several years to resolve, and new customers have been more strongly affected. All countries considered (except the United States) face the economic and political risks of using foreign launch vehicles and facilities, particularly those of the Soviet Union and China.

Despite future shuttle flights' having been reserved for noncommercial projects, U.S. satellite customers are in the best position to obtain guaranteed access to launch systems (large- and small-capacity ELVs) in the near future. Such guarantees are regulated for EOSAT; launch access for potential new customers depends on the availability of ELVs. Japan's policy of developing independent launch systems will put it on an equal footing with the United States by the mid 1990s.

France has launch agreements with ESA, on a case by case basis with NASA, and on a trial basis with the Soviet Union and China. It holds a somewhat special position within ESA because of Arianespace. Germany and Italy have access through ESA, NASA (on a case by case basis), and recently China; Great Britain's access is limited to ESA and NASA. Lack of independent access to vehicles and facilities constitutes a systematic disadvantage for these countries, but they have somewhat mitigated it by arranging for access through more than one agency.

Organization of Ground Infrastructure. Since all countries have proven ground infrastructures that are linked worldwide, no interested user is significantly disadvantaged in this area. The United States has a complex and extensive infrastructure and by far the most experience in managing national and international projects. Japan's receiving facilities, integrated into international networks, are and will be used for its own remote sensing research satellites, MOS-1 (1987) and ERS-1 (1990).

Germany, Great Britain, and Italy have available well-proven ground stations and receiving facilities through the ESA Earthnet system. In addition, Great Britain has a national infrastructure centered at Farnborough, and Italy at the ESA station in Fucino. France's ground infrastructure, in conjunction with receiving stations in Europe and overseas, is utilized for the SPOT satellites.

National Status of Satellite Technology Development. The United States leads in research, design, testing, and manufacture of scientific and commercial satellites through NASA field centers and such private firms as Hughes, Ford, and General Electric. Japan, by contrast, has little experience to date, but plans to expand efforts through NEC, Mitsubishi, and Toshiba. Japan is expected to be a competitive provider by 2000.

Europe, especially Germany, faces no barriers to developing and manufacturing high-quality satellites, perhaps even with new and innovative concepts. Germany and France have acquired experience through manufacturing research and communications satellites—MBB-ERNO and Dornier in Germany and Matra, SEP, and Aerospatiale in France. Great Britain has been a cooperative partner in manufacturing ESA satellites through IGG, Marconi, and British Aerospace; Italy has played a similar role through Selenia, Snia BPD, and Aeritalia.

National Status of Instrumentation (Sensor) Development. The United States has led in sensor technology since its early space efforts. Military development is traditionally much more advanced than civilian development, partly for financial reasons, but civilian development now receives increased attention. Current development efforts in Japan focus on SAR and optical sensors; a transition to optoelectronics is planned for the 1990s.

Most European countries face no technical barriers to instrument and sensor development. Germany has independently developed optical sensors (MOMS), France developed sensors used in the SPOT system, and Great Britain developed the microwave instrument used on ERS-1. No information on Italy is available.

Governmental Support for Satellite Technology Development. In all countries considered, R&D in satellite and sensor technology is supported by the government. U.S. support is relatively strong. Japan, Germany, and France support the development of research satellites.

Governmental Support for Data Enhancement and Value Addition. The extensive experience and lead of the United States in value addition—perpetuated by NASA's new CCDSs—can only be matched by Europe, particularly Germany, and by Japan if they increase federal support. German support is funneled through DFVLR. France provides support through CNES-financed projects and through GDTA and SPOT Image; Britain, through the National Remote Sensing Centre. Japan provides limited resources for research in evaluation and enhancement of raw satellite data, and Italy has begun to develop a data evaluation system to meet national priorities.

Governmental Support Through Exemption of Development Costs for Satellite Systems. Until now, development costs have been exempted for both operating satellite systems, Landsat and SPOT. EOSAT agreements in the United States include cost exemptions, as well as a commitment to further development and launch of a new satellite, Landsat-6. The Japanese government has assumed development costs for MOS-1 and ERS-1. Neither Germany, Great Britain, nor Italy has an indigenous commercial system. Any new system can be competitive only if development costs are assumed by the government.

Governmental Support by Risk Assumption. General subsidies for commercial utilization are not provided in any of the countries considered. Indirect subsidies, however, may be available for specific projects. For example, the U.S. government negotiates the exact amount of financial risk it will assume with the private operator, Japan has established mechanisms (e.g., the Key-TEC program) to provide limited interest-free credit, and the French government assumes some risk indirectly through its participation. No information is available on Germany, Great Britain, and Italy.

Orientation of Federal Programs Toward Eventual Commercial Utilization. The orientation toward commercialization has been strongest within U.S. and, more recently, French federal agencies. The U.S. policy of strict separation of federal and private sector task areas protects private companies, but does not provide the direct help of restricting federal agencies to basic research. The French government is unequivocal in its commercialization policy, as proven with the SPOT system.

The situation is premature in other countries. Great Britain and Italy support commercialization in theory, but have not identified it as an issue. Germany is less oriented toward commercialization, and Japan does not yet perceive it as an issue.

Data Utilization Rights. No policy differences among these countries limit commercialization. All of the countries considered here subscribe to the "open skies" policy of unrestricted access to remote sensing data.

Copyright Protection. This is a very important framework condition for data enhancers. Prior experience exists only in France and the United States, where data are protected by copyright. Raw and enhanced data are well protected in France through contracts with SPOT Image, simplifying copyright protection for data enhancers. In the United States, however, the distinction between raw and enhanced data is unclear, and enhancement and value addition create problems for EOSAT.

The situation for remote sensing is unclear in Japan. Since Germany has enacted no independent protection regulations, the conditions of suppliers are applicable. General copyright protection rules apply in Great Britain and Italy.

Government Market Regulation. Very little market regulation is practiced by any of the countries considered: the United States intervenes only when national security interests become involved. Japanese, German, French, British, and Italian markets are unregulated.

Government Sale of Data. Only the United States has wholly abandoned the sale of data. Japan sells data through RESTEC, and the Italian government handles all data sales directly. France sells Landsat data through GDTA, but has expressed a willingness to abandon this policy. Germany plans to transfer sales to the private sector, and Great Britain has negotiated to do so with Nigel Press Associates; both countries will take steps to avoid parallel sales activity.

Meeting the Government's Remote Sensing Needs. Federal agencies in all countries considered have recognized the need for complete commercialization. Negotiations with private sources are currently under way in the United States to meet government remote sensing needs.

CHAPTER

8

Satellite Remote Sensing: Conclusions

This comparison of the important prerequisites for increased commercial utilization of space for remote sensing should not be used to draw any conclusions for the political debate in Germany and in the other countries. For that purpose, the limitations of such a comparison are too great.

Here very complex issues and circumstances have been simplified to the individual criteria used for the comparison. Underlying the national space policies are often additional far-reaching interests or other forces at work (which is the situation in research) such as military needs and political issues. The intersection of political and military spheres of activity is much stronger than the commercial sphere.

Finally, information for the selected criteria is often not available because the criteria are hardly being discussed, if under consideration at all, in the national space forums in these countries. This is particularly true of Britain and, to a lesser extent, Italy, where the necessary critical mass of R&D in space has not been achieved.

The framework conditions for the individual countries are summarized below:

- **U.S.A.** The United States has excellent framework conditions for commercial utilization of space for remote sensing: space technology is well developed; indigenous launch capacity is available; the overall policies are designed to stimulate commercial utilization; hardware infrastructure is capable and efficient; the country possesses large land areas; and experience in enhancement, distribution, and application of satellite remote sensing data at more companies is better than anywhere else. Financial support from the government is now essential only for services provided in the public interest.

- **West Germany.** Although Germany possesses excellent technical know-how in sensor technology, the initiatives to commercialize this know-how have not been successful. No German firm is a shareholder in SPOT Image. Compared with the SPOT and Landsat satellites, hardware developments in Germany are not very promising for commercial applications in the near future. This means that the most promising areas for commercial utilization of satellite remote sensing are data enhancement and distribution. But here Germany lacks the dynamism found in countries with satellite operations. Also, Germany has very little long-term experience in data enhancement; the domestic data user community is insignificant, and access to the world market is difficult.

- **Japan.** The Japanese government low-key initiatives (except for launch vehicles and research satellites) are no reason to assume that the political and economic framework conditions are poor or inadequate in Japan. The objective interests of the Japanese economy tend to give low priority to satellite remote sensing, except in the special area of ocean observation. If there is an opportunity for Japanese companies in commercialization in the future, then the well-established cooperative approach of federal initiatives and coordination combined with the marketing skills and financial strength of a dynamic space industry will ensure its realization. In the long-term, the involvement of Japanese companies will probably be in the hardware segment (launch vehicles and satellite systems), rather than in the software and distribution area.

- **France.** The first and probably only European commercial organization for operation of remote sensing satellites is not coincidentally based in France. Besides technical efficiency, the unconstrained cooperation between governmental organizations and the private sector appears to be an important prerequisite in a field characterized by substantial risks and uncertainties.

- **Great Britain.** Characteristic of the framework conditions for commercial satellite remote sensing in Britain is the independence of the two major support functions, the traditional high-level scientific infrastructure (*e.g.,* National Remote Sensing Center in Farnborough) and the recently emerging private sectors for marketing and sales. The political framework conditions are characterized by the large-scale absence of a government organization, despite the complete restriction of government activities to institutions such as the NRSC. This also extends to the very limited use of federal financial resources. Even so, the economic framework conditions for enhancement and distribution of satellite data can be viewed positively, since the traditional bonds between British companies and the Third World countries will give these companies easier access to this potentially very large market for satellite remote sensing.

- **Italy.** The framework conditions in Italy have evolved mainly through the activities in the relatively large semipublic space sector. Commercialization of satellite remote sensing in Italy will have to be considered within the context of international market-based framework conditions.

The commercial utilization of space involves very high risks and uncertain profits for private industry. In the present state of commercialization, overly high expectations will certainly disappoint. First of all, technology itself is uncertain: How safe, reliable, and efficient is access to and operation in space? Is satellite remote sensing instrument technology capable of further development with respect to ground resolution? How long are the innovation cycles in instrument technology?

The political framework conditions also bear uncertainty: Will international legal regulations from an era with only two space powers be transferable to a multitude of nations in space? What cooperation policy will NASA follow in the face of growing competition from Europe and Japan, as well as from the Soviet Union, China, and other countries? Will international cooperation in space also become an instrument of industrial and trade policy, in addition to foreign policy?

With the uncertain economic framework conditions and the uncertain economic viability of commercial space utilization in general, the concerns are significant: Will federal budgets for earth observation be sufficiently large? Given the enormous number of applications of remote sensing data in the developing countries, will it be possible to transform this potential into lucrative demand?

The overall conclusion is that a new company, besides EOSAT and SPOT Image, must make a very convincing technical and economic case for commercial operation of a new satellite remote sensing system. A better system than one combining federal support with aggressive private sector marketing (similar to the model for Airbus Industrie in Europe) is difficult to visualize. Technical improvements, such as SAR for all-weather sensing and higher-resolution optoelectronic MSS, are being developed by many institutions and will hardly lead to a monopoly position for any one country.

In the other area, data enhancement and distribution, the same requirements apply for all providers:

- The existence of a large national or international market with guaranteed protection of data rights

- Further development of hardware and software for processing, analysis, and evaluation of remote sensing data

- Strong orientation toward customer needs.

For Europe, then, there is a need for demand analyses; a need for government support for the necessary software development; a need for creation of international legal protection of copyright and utilization rights covering value-added products for suppliers of raw data, as well as for users; and a need for financial support for potential customers. With these needs, the use of satellite remote sensing products and services will provide large social and overall economic benefits, but otherwise will not generate lucrative demand.

Conditions for
Commercial Utilization
of Microgravity in Space

CHAPTER

9 Background

In a narrow sense, microgravity refers to a state of near weightlessness. An object achieves this state during free fall. On Earth, near weightlessness can be realized only for very short time intervals (seconds) by free fall in drop towers or tubes, or release from aircraft or balloons. An interesting variation is "parabolic flight": the aircraft is flown in a parabolic trajectory, and during the coasting phase microgravity conditions are attained.

POTENTIAL FOR UTILIZATION

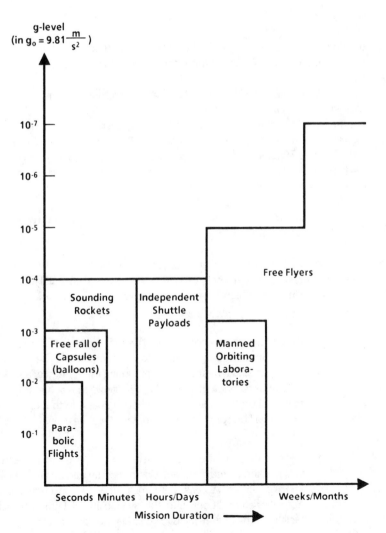

FIGURE 6.
Duration of Microgravity as a Function of g-Level for Different Kinds of Missions

For longer durations, microgravity can be attained only in a closed orbit. An intermediate step between ground-based methods and orbital flights—such as with the Space Shuttle, free flyers, or Space Station Freedom—are suborbital flights. Sounding rockets are flown in a ballistic trajectory (similar to parabolic flights) that extend into space but do not achieve Earth orbit. The duration and quality of microgravity (g-level) achievable by the above methods is shown in Figure 6.

In any discussion of microgravity, the other characteristics of space, besides weightlessness, and their effects on materials and living organisms must be considered, as well as the simulation of microgravity conditions on Earth. These are the vacuum of space and the various forms of radiation. Weightlessness itself cannot be simulated on Earth. But individual effects of weightlessness, such as containerless suspension of liquids, can be simulated (if only under severe limitations). Several effects of gravity are excluded by microgravity:

- Hydrostatic pressure in liquids

- Sedimentation in (fluid) mixtures of components with different densities

- Convective mixing in fluids.

In addition, containerless processing of liquids becomes possible: a liquid can be maintained as a suspension without the often disturbing contact with container walls.

Worldwide development activities in microgravity thus far have been almost exclusively basic academic research; industry has participated in, or conducted, microgravity experiments very infrequently. For example, the list of experiments flown on the German Spacelab-D1 mission given in Table 14 shows only 7 out of 76 experiments were proposed and conducted by private industry [84]. The objectives of the industrial experiments were basically the same as the academic experiments: they focused more on basic research rather than the development of commercial products.

A recent study in the United States [1] gives several reasons for the reservations of private industry:

- Absence of proven products. In the absence of conclusive experimental results or marketable space products, the risks involved in entry are simply too high for most private firms.

- High launch costs.

- Risk of terrestrial competition. A space R&D project requires much more time from planning to evaluation of the results than corresponding terrestrial projects. During this time, new and advanced terrestrial technologies may be developed to resolve or by-pass the technical problem being researched in space. In that case, the space project will be considered a commercial failure even if it is technically successful.

- Lack of control of access to space. In contrast to terrestrial R&D projects, interested companies do not control access to the necessary space facilities. Access still largely depends on changing political framework conditions and the availability of launch vehicles and facilities. The high risks involved in launch assurance are even more apparent with the Challenger accident in 1986, although the available

TABLE 14. *Research Fields and Experiments on the D1 Mission*

Research topic	Experiment number D1-	Experiment title	Experimenter[a]
Fluid Physics			
Capillarity	WL-FPM 04	Floating Liquid Zones	J. Da Riva, Univ. Madrid (E)
	WL-FPM 06	Adhesion Forces in Liquid Films	J.F. Padday, Kodak Ltd., Harrow (GB)
	WL-FPM 08	Liquid Motions in Partially Filled Containers	J.P.B. Vreeburg, NLR, Amsterdam (NL)
Marangoni Effects	PK-HOL 03	Bubble Motions Induced by Temperature Gradients	D. Neuhaus, DFVLR, Köln (D)
	PK-MKB 00	Marangoni Convection in an Open Vessel	D. Schwabe, Univ. Giesen (D)
	WL-FPM 07	Marangoni Flows	L. Napolitano, Univ. Naples (I)
	WL-FPM 01	Marangoni Convection in Gas/Liquid Mass Transfer	A.A.H. Drinkenburg, Univ. Gronigen (NL)
	WL-FPM 05	Thermocapillary Movements Under Microgravity	J.C. Legros, Univ. Brussels (B)
	PK-HOL 01	Chemical Wave Transport	A. Brewersdorff, DFVLR, Köln (D)
Diffusion	WL-HTT 00	Interdiffusion in Molten Metals	K.H. Kraatz, Tech. Univ. Berlin (D)
	WL-GHF 01	Thermal Diffusion/Soret Effect	J. Dupuy, Univ. Lyon (F)
	PK-IDS 00	Interdiffusion in Molten Salts	W. Merkens, Tech. Univ. Aachen (D)
	WL-IHF 05	Homogeneity of Glasses	G.H. Frischat, Tech. Univ. Clausthal (D)
	ME-GPRF 2	Diffusion of Liquid Zn and Pb	R.B. Pond, Marvalaud, Inc. (USA)
	WL-GHF 07	Thermomigration of Co in Sn	J.P. Praizey, CEN, Grenoble (F)
	ME-SAAL	Containerless Melting of Glass	D.E. Day, Univ. Missouri-Rolla (USA)
Critical Point Phenomena	MD-HPT 00	Heat Capacity at Critical Point	J. Straub, Tech. Univ. Munich (D)
	PK-HOL 02	Density Distribution and Phase Separation at the Critical Point	H. Klein, DFVLR, Köln (D)
Solidification			
Solidification Front Dynamics	PK-HOL Q4	Boundary Layers in Solidification of Transparent Melts	A. Ecker, Tech. Univ. Aachen (D)
	MD-GFO Q1	Phase Boundary Surface Diffusion	H.M. Tensi, Tech. Univ. Munich (D)
	MD-GFO 02	Convection in Solidification	S. Rex, Tech. Univ. Aachen (D)
	WL-GHF 02	Cellular Morphologies in Pb-Tl Alloys	B.B. Billia, Univ. Marseille (F)
	WL-GHF 04	Dendritic Solidification of Al-Cu Alloys	D. Camel, CEN, Grenoble (F)
	MD-ELI 04	Directional Solidification of InSb-NiSb Eutectics	G. Müller, Univ. Erlangen-Nürnberg (D)
	WL-IHF 09	Nucleation of Eutectic Alloys	Y. Malmejac, CEN, Grenoble (F)

TABLE 14. *Continued*

Research topic	Experiment number D1-	Experiment title	Experimenter[a]
Solidification (*continued*)			
Crystal Growth	WL-MHF 01	Float Zone Growth of Silicon	A. Cröll, Univ. Freiburg (D)
	WL-MHF 04	Crystallization of an Si Sphere	H. Kölker, Wacker Chemie, Munich (D)
	WL-GHF 03	Growth Rate Measurement of Doped InSb	C. Potard, CEN, Grenoble (F)
	MD-ELI 01	III-V Compound Semiconductors	K.W. Benz, Univ. Stuttgart (D)
	WL-MHF 02	THM Growth of GaSb	K.W. Benz, Univ. Stuttgart (D)
	WL-MHF 03	THM Growth of CdTe	R. Schönholz, Univ. Freiburg (D)
	MD-ELI 02	THM Growth of PbSnTe	M. Harr, Battelle, Frankfurt (D)
	MD-ELI 03	Gas Phase Crystallization of CdTe	M. Bruder, Univ. Freiburg (D)
	WL-GHF 05	Ge/GeI_4 Chemical Growth	J.C. Launay, Univ. Bordeaux (F)
	WL-GHF 06	Ge-I_2 Vapor Phase Growth	J.C. Launay, Univ. Bordeaux (F)
	ME-GPRF 4	Vapor Growth of Alloy-Type Semiconductor Crystals	H. Wiedemeier, Rensselaer Polytechnic, Troy (USA)
	ME-GPRF 5	Semiconductor Materials Growth in Low Gravity	R.K. Crouch, NASA Langley (USA)
	WL-CRY 00	Protein Crystal Growth	W. Littke, Univ. Freiburg (D)
Composites	WL-IHF 01	Separation of Immiscible Melts	H. Ahlborn, Univ. Hamburg (D)
	WL-IHF 02	Alumina Particles Preceding a Copper Solidification Front	J. Pötschke, Krupp Research (D)
	WL-IHF 06	Particle Behavior in Melt and Solidification Fronts	D. Langbein, Battelle, Frankfurt (D)
	WL-IHF 08	Melting and Solidification of Metallic Composites	A. Deruyttere, Univ. Leuven (B)
	WL-FPM 03	Mixing and Demixing of Transparent Fluids	D. Langbein, Battelle, Frankfurt (D)
	WL-FPM 02	Separation of Liquid Phases and Bubble Dynamics	R. Nähle, DFVLR, Köln (D)
	ME-GPRF 3	Liquid Phase Miscibility Gap Materials	H.S. Gelles, Columbus (USA)
	WL-IHF 04	Ostwald Ripening in Metallic Melts	H. Fischmeister, MPI, Stuttgart (D)
	WL-IHF 03	Support Skin Technology	H. Sprenger, MAN, Munich (D)
	WL-IHF 07	Skin Casting of Cast Iron	H. Sprenger, MAN, Munich (D)
Biology	BR 32/33 CH	Effect of Spaceflight on Lymphocyte Proliferation Cell Proliferation	A. Cogoli, ETH, Zürich (CH)
	BR 21 F		H. Planel, Univ. Toulouse (F)
	BR 48 F	Mammalian Cell Polarization	M. Bouteille, Univ. Paris (F)
	BR 27 D	Circadian Rhythm	D. Mergenhagen, Univ. Hamburg (D)
	BR 58 F	Antibacterial Activity	R. Tixador, Univ. Toulouse (F)
	BR 28 D	Growth and Differentiation of Bacillus Subtillis	H.D. Mennigmann, Univ. Frankfurt (D)
	BR 07 I	Effect of Microgravity on the Mechanism for Genetic Recombination	O. Ciferi, Univ. Pavia (I)
	BR 16 D	Contraction Behavior and Protoplasma Flow	V. Sobick, DFVLR, Köln (D)
	BR 19 D	Dosimetric Mapping in BIORACK	H. Bucker, DFVLR, Köln (D)

TABLE 14. *Continued*

Research topic		Experiment number D1-	Experiment title	Experimenter[a]
Biology *(continued)*	Development Processes	BW-STA 00	Embryonic Development of Vertebrate Gravity Receptors	J. Neubert, DFVLR, Köln (D)
		BR 52 NL	Determination of the Dorsoventral Axis	G. Ubbels, Univ. Utrecht (NL)
		BR 15 E	Distribution of Cytoplasmic Determinants	R. Marco, Univ. Madrid (E)
		BR 18 D	Embryogenesis and Organogenesis	H. Bucker, DFVLR, Köln (D)
	Plants	BW-BOT 01	Graviperception	D. Volkmann, Univ. Bonn (D)
		BW-BOT 02	Geotropicalism	J. Gross, Univ. Tübingen (D)
		BR 39 F	Perception of Gravity in the Lentil Root	G. Perbal, Univ. Paris (F)
		BW-BOT 03	Induction of Somatic Embryogenesis	R.R. Theimer, Univ. Munich (D)
Medicine	Function of Gravity on Perceived Organs	VS-ES 201	Reaction of a Human Vestibular Apparatus Under Space Flight Conditions	R. von Baumgarten, Univ. Mainz (D)
		VS-NS 102	Vestibular Adaptation	L. Young, MIT, Cambridge (USA)
	Adaptation Process	BW-ZYO 00	Venous Pressure Measurements in Space	K. Kirsch, Tech. Univ. Berlin (D)
		BW-TOM 00	Measurement of Intraocular Pressure Under Microgravity	J. Draeger, Univ. Hamburg (D)
		BW-BIM 300	Body Impedance Measurement	F. Baisch, DFVLR, Köln (D)
	Cognitive Behavior Under Microgravity	ROS 230	Mass Discrimination	H.F. Ross, Univ. Stirling (GB)
		LAN 200 SDS	Spatial Discrimination in Space	A.D. Friederici, MPI, Nijmegan (NL)
		LAN 200 GPS	Gesture and Speech in Microgravity	A.D. Friederici, MPI, Nijmegan (NL)
		JUF 250	Reaction Time Measurements	M. Moschek/J. Hund, Mühltal (D)
	Navigation	NX-UST 00	Ear Synchronization	S. Sterner, DFVLR, Oberpfaffenhofen (D)
		NX-EWE 00	Distance Measurements	D. Richter, SEL, Stuttgart (D)

Legend:
B = Belgium
CH = Switzerland
D = Federal Republic of Germany
E = Spain
F = France
GB = Great Britain
I = Italy
NL = Netherlands
USA = United States of America

worldwide space transportation fleet is now increasingly capable of meeting all projected launch needs.

- Markets for space products have not developed, so the financial payoff from space projects is difficult to project. The willingness to invest in such projects is correspondingly small.

- Lack of knowledge of the possibilities offered by microgravity.

- Second and third party liability for damage to the Shuttle.

- Long lead times resulting in delayed return on investments.

In every country that conducts basic research in space, there are intensive efforts to overcome private sector reluctance and to involve industry in utilization of space. Potential commercial applications in microgravity are expected mainly in finely dispersed alloys, ceramic materials, single crystals of semiconductor materials, and pharmaceuticals [1,9,85]. However, these applications probably will not occur until early in the next century after a period of basic research and technology development, including the early years of Space Station Freedom's operation [86].

In particular, growth of large, high-quality single crystals is indispensable for applications in optics, high-frequency technologies, and micro- and optoelectronics. Table 15 lists some applications for synthetic single crystals [87]. Because of the large number of such applications, industrial interest in crystal growth of semiconductor materials in microgravity is expected to increase significantly [88,89]. Also, space-based growth of protein crystals is becoming a key business development area; major U.S. and international pharmaceutical companies have sponsored protein crystal growth experiments on the U.S. Space Shuttle, the Soviet MIR space station, and Chinese unmanned spacecraft [90].

Independent of the above mentioned classes of materials, activities in space materials processing can have various objectives, including the production of materials in space, research on products and processes for future production in space, and research to improve materials and processes on Earth and to produce ideal material standards. For the present and near future, manufacturing in space is unlikely, except perhaps of unique samples of highly perfect single crystals or small quantities of very rare (and expensive) pharmaceutical products (high value-added products). Thus, the emphasis in utilizing microgravity will mainly be in the third area. Microgravity allows the study of basic processes under idealized boundary conditions and to understand the effect of gravity on these processes. Such understanding can be applied to improve materials processing on Earth.

For example, it has been demonstrated by growing silicon crystals in space that the inhomogeneities (striations caused by uneven dopant distribution) in Earth-grown crystals can be attributed to surface-tension driven flows (the Marangoni effect), rather than gravity-driven flows. On Earth, these inhomogeneities can now be eliminated by covering the "free" surface of the melt, from which the crystal grows, with a thin oxide film [91].

Materials produced under the ideal conditions of space can be analyzed on Earth to provide new information about their properties. For example, proteins and other complex substances sometimes can be crystallized more easily under microgravity. These higher-quality crystals, in turn, permit more exact analysis of the crystal structure of the proteins on Earth.

Space-produced optimal material samples also serve as ideal standards because they exhibit the maximum achievable improvements in the material under ideal boundary conditions. Thus, the space samples provide important reference points in setting goals for advanced material developments on Earth.

In the following sections, the status of space activities related to microgravity utilization is briefly reviewed for the countries considered. Because of the division of space activities in Europe, the activities of the European Space Agency will be reviewed in a separate section from the individual European countries.

INTERNATIONAL STATUS OF DEVELOPMENT

TABLE 15. *Examples of Applications for Synthetic Single Crystals*

Property and application	Types of crystals
Semiconductor materials (electronic components, integrated circuits)	Silicon, germanium, gallium arsenide, indium antimonide
Lasers, masers (source for coherent EM radiation)	Ruby, Nd-doped calcium tungstate, Nd-YAG, Nd-YAP, neodymium pentaphosphate, gallium arsenide
Microwave technology (emitters, frequency stabilizers, power limiters)	Gallium arsenide, yttrium ferrite, YIG
Solar cells (conversion of EM radiation into electrtical energy)	Silicon (single crystal films), selenium
Optoelectronic LED (conversion of electrical energy into EM radiation)	Gallium arsenide, gallium phosphide
Electroacoustics, high frequency technology (frequency stabilization of transmitters, tunable ultrasonic generators, control of precise timers)	Quartz, KNT, ADP, lithium tantalate
Optics (polarizers, UV and IR transparent lenses, filters, prisms, achromators, high-temperature optics)	Calcite, sodium nitrite, fluorite (CaF_2), alkali halides, magnesium oxide, alum, copper sulfate, quartz
Nonlinear optics (light modulation, opening new wavelengths and wavelength ranges for coherent radiation by frequency doubling and frequency mixing)	KDP, lithium niobate, silver gallium arsenide, proustite
Magnetic optics (light modulation)	YIG, iron borate
Dielectric properties (ferro, pyro, and piezoelectric properties)	Barium titanate, TGS, KDP, quartz
Optical (holographic) storage	Lithium niobate, stilbene derivatives
Magnetic bubble storage	YIG (single crystal films on GGG)
Radiation detectors (scintillation counters, IR detectors)	Thallium-activated sodium iodide, anthracene-activated naphthalene, lead, tin, telluride
Luminescence, photoconductors	Zinc sulfide and selenide, cadmium sulfide
Monochromators (selectors for x-rays)	Lithium fluoride, quartz, graphite, aluminum
Heat-conducting high-strength materials (aircraft gas turbines)	Metallic crystals
Hardness (grinding, polishing, cutting, sawing, drilling rigs, drawing dies)	Corundum, carborundum, diamond
Isolating materials	Mica
Gems	Ruby, sapphire, alexandrite, smaragdite

Because of its effect on the status and the continued utilization of microgravity, particularly on the initial efforts of private companies to participate in microgravity activities, the complete loss of a major retrievable launch system for microgravity payloads—all Shuttle flights were halted after the Challenger accident in January 1986—has had important ramifications. Since the consequences of the Challenger accident are not limited to the United States, they also will be discussed in a separate section.

West Germany

Within the German space programs (1988 budget, DM 1.2 billion), the "utilization of weightlessness for analysis of scientific and long-term technological issues" [42,92] is a major thrust: the 1988 budget for materials processing was approximately DM 149 million, of which DM 36 million was earmarked for ESA microgravity programs [44]. This is reflected in German participation in international microgravity programs and in its development and implementation of an independent national program.

The three large international (ESA) programs—Spacelab-1, the free-flying platform EURECA, and the Columbus project (the Free-Flying Laboratory and the Polar Platform are the European contributions to Space Station Freedom)—are Germany's major space activities. In all three programs, Germany provides the largest share of the total funding (Spacelab [FSLP], 56 percent; EURECA, 54 percent; Columbus, 38 percent), and the German space industry leads these programs. In accordance with the share of each country participating in the ESA programs, Germany also has the largest share in the utilization of Spacelab and EURECA.

The large national programs are the Spacelab missions D1 (completed in October 1985 on Shuttle mission 22) and D2 (scheduled for launch in December 1991 on Shuttle mission 52 [27]), the TEXUS sounding rocket and the MAUS programs, the SPAS and Astro-SPAS missions, and the Raumcourier recoverable capsule. In addition, Germany has flown one mission on a Chinese Long March rocket and is considering use of additional Chinese and Soviet flight opportunities to conduct microgravity research [90,93]. These and other programs are described individually in Chapter 13 under "Basic Types of Experimental Options under Microgravity."

With this strong emphasis, Germany has achieved the leading position in microgravity activities in Europe. Currently, the country possesses what is probably the largest base of space-related experimental know-how, compared even with the United States. This is the result of a large number of experiments carried out on the above missions, the wide variety of the experimental goals and technical designs, and the number of investigators. The investigators are, as mentioned earlier, primarily from universities and scientific organizations, and the few experiments conducted by industrial investigators were not commercially oriented.

France

France spent FFr 5.87 billion in 1987 for civilian space projects [47], but the emphasis is in the "classic" fields—launch vehicles, telecommunications, and satellite remote sensing—rather than microgravity. Only the general space R&D program (FFR 191 million, about 3 percent of the total 1986 R&D budget) includes funding for microgravity research. The space budget is divided as follows [53]:

- 25 percent for science and basic technology
- 22 percent for Earth observation

- 20 percent for telecommunications

- 17 percent for launch vehicle technology

- 16 percent for on-orbit infrastructure.

Microgravity activities are included only in the areas of science and basic technology and on-orbit infrastructure. In 1985, CNES recognized that microgravity research had been sidelined and it planned to increase its support for microgravity [49]:

> *A laboratory for research in the field of microgravity is one area of space research which CNES should support and develop on a small scale. The number of scientists who work in space research should be increased to about two hundred.*

France does not have an independent microgravity program and it does not offer any dedicated missions under French direction. Despite this, France is familiar with and contributes to developments in microgravity by conducting materials science and life science experiments within international research programs. This participation is in ESA programs and in the national programs of other ESA members, as well as in bilateral projects with the United States and the Soviet Union. Examples of French participation in other missions are the 12 experiments on the German D1 mission (nearly 16 percent of the total number), experiments on TEXUS, experiments on the NASA Spacelab-3 mission, experiments planned for EURECA-1, and experiments on the Soviet MIR space station and unmanned retrievable capsules as part of the joint French-Soviet agreements, as well as with China.

Universities and research centers like Centre d'Étude Nucléaires de Grenoble (CENG), Laboratoire d'Étude de la Solidification (LES), Centre National de la Recherche Scientifique (CNRS), and others conduct a wide variety of research in microgravity with support from CNES [51]. Industrial participation in microgravity research has been very low in France, although this situation is changing. A number of chemical companies are planning to conduct large-scale experiments in protein crystallization and to construct a special experimental facility [94]. This facility will be used on board the Soviet MIR space station as part of the French-Soviet cooperation in space, but talks have been held with INTOSPACE in Germany for possible inclusion of the facility in the D2 mission [93].

With these bilateral projects, especially cooperation with the Soviet Union, France has assumed a special position among ESA member states. France has made a determined effort for autonomy in space from the United States and also from its other partners within ESA.

Italy

In Italy, spending for space has increased significantly during the last few years. Until 1982, the annual national space budget was about L 100 billion, but it was increased to L 362 billion in 1985, and increases of L 700 billion annually are planned to the year 2000. This increase is directly related to Italian participation in the ESA Columbus project. With a 25 percent share of Columbus, Italy has expressed its increasing interest in microgravity. Of the above budget figures, nearly 15 percent is for satellite remote sensing, 5 percent for microgravity (without Columbus), and 25 percent is the Italian contribution to Columbus. The remaining 42 percent is used for telecommunication and for contributions to other ESA programs.

The utilization of microgravity has thus far been insignificant in Italy; it plans no independent mission and has recently initiated its first project outside ESA. However, a bilateral German-Italian project TOPAS (explained in Chapter 13 under "Structure of the. . .Experimental Options") has been initiated to place microgravity payloads in Earth orbit using Scout rockets. Within ESA, Italy has participated in utilization of Spacelab (Spacelab-1 and D1) and Italian experiments have been proposed for EURECA-1 and Spacelab-D2. The above involvement in the Columbus program, as well as in the founding of INTOSPACE by Aeritalia and MBB-ERNO (described in Chapter 13 under "Accountability and Oversight") suggest, however, that Italy will be more involved in microgravity in the future. A Microgravity Advanced Research and User Support Center (MARS) has been established at the University of Naples to provide technical assistance and consultation for microgravity users [95].

Great Britain

Britain spends slightly more than £ 100 million per year for space science and civil space R&D [57]. Of that, nearly 75 percent is spent on ESA contributions. This means that Britain, like Italy, is focusing its space activities on participation in international programs with less emphasis on national programs. British space activities are concentrated in communications satellites (nearly 47 percent of the funding), but satellite remote sensing is expected to grow. The situation for microgravity is comparable to that in Italy. Whether microgravity will increase in the future, initiated perhaps by the British National Space Centre (BNSC) established in 1985, is not foreseeable at this time. Britain has a 15 percent share in the Columbus program.

European Space Agency

Fifteen countries (including Canada and other associated countries, see Table 9) share a part of their space activities within ESA. Two kinds of space projects are carried out in ESA, mandatory and optional projects. The mandatory projects are financed from the mandatory contributions of all member states based on their respective GNPs. The optional projects, on the other hand, are selected by choice for participation by the member countries, which also provides each country the opportunity to emphasize certain themes.

With the establishment of ESA, the member countries now have a common space policy based on the following principles:

- Encourage European cooperation in space
- Utilize the available resources of member countries—test facilities, mission control centers, and ground stations—for joint projects
- Improve the competitiveness of the European space industry
- Promote equitable financial return, so that contract awards to industries in a member country are in direct relation to the financial contributions of that country
- Make jointly developed technologies available to all member countries
- Convert previously developed space technologies into operational systems for use by member countries
- Assist in operating these systems in the member countries

In the area of microgravity utilization, ESA first participated in the maiden Spacelab flight SL-1 (FSLP) in 1983. This was followed by participation in the German Spacelab-D1 mission in 1985. The combined number of microgravity experiments flown on SL-1 and D1 was equal to the total number of U.S. experiments on 21 Shuttle flights [96]. Other large projects are also being prepared:

- **EURECA.** Design, construction, and flight of a free-flying platform (see Chapter 13, "Basic Types of Experimental Options under Microgravity")

- **Columbus.** European contribution to the planned Space Station Freedom, including the Columbus Attached Laboratory and the Columbus Free-Flying Laboratory.

ESA also participates in the national programs of its members, for example, the German TEXUS and Spacelab-D2 projects (20 percent ESA share). This participation is mainly in the form of providing its own payloads and experimental facilities. To provide technical support to microgravity users, ESA has established a Microgravity User Support Center (MUSC) (described in Chapter 13 under "Institutions Providing...Support"), at the DFVLR Institute for Space Simulation in Cologne, Germany.

ESA's projected spending for microgravity projects, based on its long-term plan (Table 8), includes microgravity-related expenditures in the general budget [69]. Table 16 lists the national shares of the largest optional ESA microgravity projects. In all these projects, Germany has the largest share and the leadship role [48].

ESA is also making efforts to encourage the commercial utilization of space. For this purpose, the Commercialization Office was established in 1985 in Paris [97].

TABLE 16. *Contributions of the Countries Considered to Key ESA Optional Microgravity Programs (percent)*

Programs	1	2	3	4	5
Germany	27.57	31.43	56.27	53.66	38.00
France	15.50	15.73	12.81	17.31	15.00
Italy	7.50	17.00	4.41	17.33	25.00
United Kingdom	1.35	1.34	8.44	2.10	15.00

Legend: 1 = Microgravity Phase 1   4 = EURECA
         2 = Microgravity Phase 2   5 = Columbus Development Program
         3 = Spacelab FSLP

Japan

Japan recognized very early the potential of space [80]. As a result, the space budget increased strongly in the 1970s and was approximately ¥ 141,780 million for 1988 [80]. Thus far, the emphasis in Japanese space activities has not been in microgravity, but in launch systems and communications satellites. Expenditures for microgravity in 1987 were only about 5 percent of the total space expenditures. Until now, very few Japanese experiments have been conducted under microgravity. Since 1973, Japan has carried out experiments on Skylab and the Shuttle, the European Spacelab, and Japanese sounding rockets. However, it is Japan's declared goal to exploit microgravity more extensively in the future. This goal is reinforced by four key Japanese decisions [80]: to contribute its own pressurized module (JEM) to the Space Station Freedom; to perform experiments on foreign Spacelab missions (German D2 mission and IML-1, described in Chapter 13 under "Basic Types of Experimental Options"); to fly a dedicated Spacelab mission, SL-J, which will carry the First Material Processing Test (FMPT) payload [98], on Shuttle mission 48 [27]; and to develop an independent unmanned free-flying research platform.

Noteworthy among the Japanese microgravity activities is the strong involvement of private industry [99,100]:

- Presently five groups of Japanese companies (totalling 300 firms) are studying the commercial utilization of Space Station Freedom.

- The Japan Space Utilization Promotion Center (JSUP) was established in February 1986 with the cooperation of 41 independent firms representing such diverse disciplines as space technology, materials science, and pharmaceutical technology. JSUP's foremost goals stress in-depth microgravity research, and it manages all activities associated with participation in the IML missions.

- In April 1986, the Space Technology Corporation (STC) was established by the Japanese Key Technology Center (Key-TEC) and six industrial firms—Ishikawajima-Harima Heavy Industries, Toshiba Corporation, NEC Corporation, Hitachi, Fujitsu, and Mitsubishi Electric Corporation. The initial capital is ¥ 107 million with a government share (through Key-TEC) of 70 percent. These firms are also developing their own individual experiments for the planned German Spacelab-D2 mission through STC, as well as for participation in the TEXUS program.

- The Institute for Unmanned Space Experiments with Free Flyers (USEF) was established in May 1986 to develop the technology for the planned free-flying platform and to promote the use of this unmanned system for space experiments. Thirteen companies (including Mitsubishi Electric Corporation, NEC, Toshiba, and Hitachi) provided the starting capital of nearly ¥ 100 million.

Japan still lags behind the United States and Germany in its scientific knowledge of microgravity and in experiment technology. But Japan is making a vigorous effort to close this gap. Japanese institutions (through NASDA) are contracting for critical analyses of the microgravity experiments conducted in the United States and Germany, thereby "purchasing" the know-how developed through these experiments [101]. Japanese experimenters are attempting to develop contacts and cooperation with experienced investigators from the United States and Europe through bilateral cooperative and information exchange agreements and microgravity seminars [102,103]. Thus, a paradoxical situation has developed in which the United States and Europe have a wide lead over Japan in acquiring know-how, yet Japan is ahead in processing, evaluating, and using this know-how for future missions.

United States of America

Microgravity research in the United States started in the late 1960s with simple demonstration experiments in solidification, fluid physics, and electrophoresis on the Apollo spacecraft. These activities were continued during the Skylab

(1973–1974) and the Apollo-Soyuz (1975) programs [1]. After Apollo, microgravity experiments in Earth orbit were interrupted until the first Space Shuttle flights. During this interruption, NASA continued its microgravity experiments in a suborbital rocket program, Space Processing Applications Rocket (SPAR), in drop tubes, or using aircraft flying in parabolic trajectory. These programs continue, not in the least because of the temporary halt in Shuttle operations [104]. Other orbital microgravity programs based on the Shuttle were two Spacelab missions and the Get Away Special (GAS) program, as well as future use of the middeck augmentation module, Spacehab [74]. In addition, future microgravity experiments could be carried out on the planned automated free flyers ISF and CDSF and on proposed facilities employing the Shuttle's external fuel tanks (Space Phoenix, TDP) [9,74]. Similarly, unmanned retrievable reentry capsules are also planned for microgravity experiments [9]. Finally, activity is under way to prepare industrial experiments for the planned U.S. Microgravity Laboratory (USML-1) mission, scheduled for launch in March 1992 [27,104], as well as to define industrial user requirements for microgravity experiments aboard Space Station Freedom [105]. Details of these programs and facilities are provided in Chapter 13 under "Basic Types of Experimental Options."

On an absolute basis, even in the United States, private industry is still very reluctant to invest in microgravity activities, although some of the largest U.S. industrial corporations are involved in commercial space activities related to microgravity utilization [104]. Industrial involvement is clearly greater than in Europe and Japan. Presently, some 120 companies are either actively involved in or have expressed an interest in microgravity research [104,106].

Like the respective European and Japanese agencies, NASA is very interested in commercializing microgravity utilization and encourages it in the following ways:

- The Office of Commercial Programs (Code C) was established in 1984 within NASA.

- The Commercial Space Policy and Implementation (CSPI) plan and the Commercial Space Initiative have been implemented.

- Centers of Excellence and Centers for the Commercial Development of Space (CCDS) have been established.

- The ELV industry has been transferred to the private sector.

NASA hopes that these steps will stimulate private investment and activity in microgravity utilization that, in turn, will return economic benefits to U.S. industries, advance science and technology, help maintain U.S. space leadership, and enhance the nation's competitive position in international trade, thereby improving the U.S. balance of payments.

Consequences of the Space Shuttle Challenger Accident

The tragic accident of the Space Shuttle Challenger on January 28, 1986, had major consequences for short-term commercial space activities and for industrial involvement in microgravity research.

- Operation of the U.S. manned space transportation system (STS) was halted until September 29, 1988, when the Shuttle flights resumed with the launch of Discovery.

- Orbital missions for microgravity research in all countries considered are designed specifically for the Shuttle, so all experiments requiring microgravity for longer than a few minutes could not be performed during the Shuttle hiatus.

- U.S. military payloads were given priority over U.S. civilian and foreign payloads when Shuttle flights resumed. Military payloads will increasingly be launched on ELVs.

- The accident has placed the technical reliability of a microgravity scenario in doubt. As discussed in Chapter 11, reliability and cost accountability of a space program are important requirements for the commercial user. Skepticism in the industry has increased. (The delay in the commercialization of microgravity caused by the Challenger accident is also discussed under "Demand for Utilization of Microgravity Capabilities," below.)

- In the United States, companies already involved in space postponed their plans in microgravity. The experimental plans of McDonnell-Douglas (MDAC) and 3M were hardest hit. Both companies, as well as other U.S. companies, are still very interested in microgravity research, but are seeking ground-based solutions as well as alternate flight opportunities.

- The interruption of large orbital missions has led to a partial loss of previously acquired technology (discussed more fully in Chapter 13 under "Development of Basic Technology").

- Independence from the United States STS became possible for European and Japanese users. Despite past successful U.S.-European and Japanese-American cooperation, the question of political guarantees for access to space is still unresolved.

- As discussed in Chapter 13 under "Structure of . . . Experimental Options," alternatives to the U.S. STS are being sought and European efforts for autonomy in launch systems are increasing. These efforts were under way before the Challenger accident, but have increased as a result of the accident. The European efforts are aimed at

  —Developing an independent manned retrievable transport system, Hermes, to be launched by Ariane-5. Other transportation systems being considered are the German Sänger as a fully reusable two-part system [107,132] and the British HOTOL as a single-stage-to-orbit launch vehicle [108]. Although initial funding for Hermes has been approved by ESA, the program will be subjected to a progress review after three years [109].

  —Cooperating with the Soviet Union and China.

  —Developing an independent retrievable system, such as Raumcourier and Revex, to return microgravity payloads to Earth.

- The Presidential directive of February 1988 returned the United States to a mixed fleet of space transportation vehicles, so the country would never again be entirely dependent on a single vehicle for access to space.

The size and growth of future demand for microgravity are difficult to predict, since these depend on the availability of flight opportunities: more results from space experiments will generate greater interest and hence greater demand. This is analogous to other scientific fields in the initial stages of development (*e.g.*, low-temperature physics). Microgravity research will be needed far into the future, as evidenced by the increasing positive reaction to requests for proposals for space missions. For example, over 200 proposals were submitted for the planned German Spacelab-D2 mission, of which only about a third might be accepted [44].

A distinction must be made between publicly and privately funded demand. In 1981, Waltz [110] projected the development of both these demands in the future. This projection (Figure 7) is still useful.

That demand is dependent on available missions is clearly evident (Figure 7). When the Shuttle program (which offers more regular flight opportunities than previous U.S. programs) became operational (1980–1981), demand began to grow (bottom of figure). It is assumed that this growth is entirely financed by the private sector, while government-sponsored demand remains constant. This projection is consistent with the common belief that, in the short term,

DEMAND FOR UTILIZATION OF MICROGRAVITY CAPABILITIES

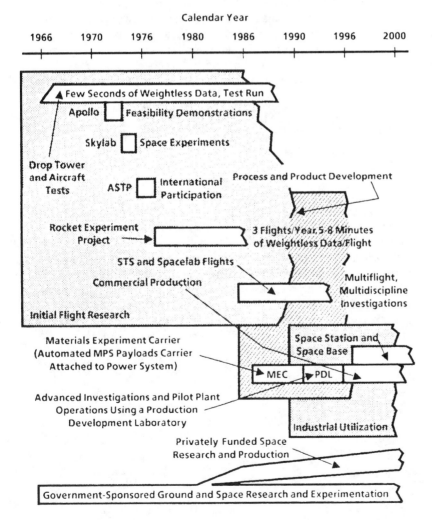

FIGURE 7.

Projected Development of Government- and Privately Funded Demand for Microgravity Utilization as a Function of Flight Opportunities

government-sponsored space research will stimulate other private research and innovation [14,111].

With respect to private financial investment, Figure 7 should be revised to include the recent delays (mainly at the start of the Shuttle/Spacelab program and the Challenger accident), which total nearly five years. Based on a revised projection, the share of privately funded research will nearly equal the government share by about 1995. Even this projection appears too optimistic under current conditions: most recent opinion suggests production in space is unlikely until early in the 21st century [86]. Industrial involvement will be limited to the manufacture of calibration and pure standards by taking advantage of ultrapure processing (the absence of wall contamination allowed by containerless processing), and undisturbed processing (the absence of thermal convection and sedimentation). It should be possible to determine with little time and effort whether additional terrestrial efforts are needed to optimize the product or the manufacturing process further (compare the section "Potential for Utilization").

Experts have frequently expressed the opinion that demand is insufficient to utilize fully the capacity of the planned Space Station Freedom. This opinion, however, overlooks two factors:

- If the initial equipment aboard S.S. Freedom is designed as much as possible for pursuing high-value, low-volume research areas as in a standard terrestrial laboratory (i.e., storage of material, processing and analysis facilities), fast and flexible reactions to commercial "orders" will be possible on board S.S. Freedom. Materials and equipment transportation to and from S.S. Freedom will have to be improved.

- In the short term, demand will be stimulated by providing more flight opportunities. Here Germany has a unique advantage over all other partners (with the exception of the Soviet Union with its operating MIR space station): it has an operational and very reliable flight program, TEXUS.

The authors' own assessments [112] of the growth in demand for experimental uses of Space Station Freedom (only the European Columbus Attached Laboratory was considered) indicate that demand could increase until the year 2000 to an even greater level than that currently projected by ESA [113]. This increase will arise mainly from more intensive efforts to explore the potential of microgravity by academic researchers and the private industry. Should these research efforts, or even actual processing under microgravity, be successful, the demand will increase further, especially on automated, unmanned platforms. On the other hand, without any commercial utilization, demand will decrease to a point at which the need for a permanent space station will be in question.

CHAPTER

10 Commercialization Foundations and Requirements

FUNDAMENTAL
CONSIDERATIONS
BASED ON A
TYPICAL
EXPERIMENT
SCENARIO

The basic considerations of supportive and disruptive forces in the innovation process, discussed in Part I on satellite remote sensing, also apply to a limited extent to the microgravity segment. The differences are mainly in the relative size of the respective projects. Whereas a remote sensing project is a relatively comprehensive venture, including all stages from data generation to processing and distribution, entry into a microgravity venture is technically fairly simple. Depending on the user, the project can range from experiments with a single sample to the flight of an entire laboratory facility. The typical steps of an industrial experimental project, largely independent of the size of the project, are given below:

- Identify existing production problems and discuss potential solutions, including use of weightlessness.

- Analyze available data and consult with technical experts. Enter into a research partnership.

- Develop experimental ideas and project conception.

- Decide to execute the project.

- Plan the microgravity experiment.

- Complete preparative work and accompanying laboratory research.

- Convert the terrestrial laboratory-based experiment design to a space-qualified version.

- Perform the microgravity experiments.

- Evaluate the results.

- Exploit the results.

At each of these steps, it is possible to interrupt the project, which makes it easier to reach a decision for execution or termination of a space project at any time.

The different framework conditions have a positive or negative effect on the various phases of the project. To derive the requirements of each phase of the project, the scope of the activities in each phase is described below.

Steps Prior to Idea Generation and Project Conception

The evaluation of ideas for space experiments within a private company assumes a positive attitude towards space, in general, and towards microgravity utilization, in particular. At both the technical and decision-making levels of a company, there must exist a willingness to discuss the scientific and technical feasibility, as well as the commercial potential of space.* A general interest in space is fostered by several factors in the company's environment:

- The importance attached to and the media treatment of space activities in the country

- The political climate created through clear federal goals and priorities

- The significance of microgravity within national space programs

- Successes of previous microgravity activities (which generate added publicity for microgravity).

The most positive influence, however, may come from proven commercial successes of space utilization of the type not yet achieved anywhere in the world. With the exception of initial exploratory experiments within the Skylab program, significant microgravity research began in the early 1980s. The absence of space products will have a negative effect on the general attitude of the private sector only if, under pressure for rapid commercialization and quick payback, premature publicity generates an expectation level that cannot be met by the present state of microgravity development.

The results of basic academic research in microgravity also have a public influence. They are judged successful if they receive recognition from the scientific community.

The authors' own discussions with industry representatives have shown that there is a positive attitude toward and interest in microgravity within the private sector; other studies agree [86,114]. Yet very few companies had specific ideas about what they themselves could do in space. Such ideas must be generated within the company itself. In the long term, ideas and results generated by ongoing basic research in a consistent manner will be adopted and used by industry for commercial purposes. In the short term, however, ideas for microgravity research will probably have to evolve from current R&D activities within the company. Private firms must, therefore, have technical advocates on staff who can evaluate the potential of microgravity in relation to the commercial interests of the company. They must also receive effective support through close contact with academic microgravity research, for example, in joint projects with universities. This support includes dissemination of information about the physical and chemical effects of microgravity—that is, direct involvement of industrial researchers in studying basic scientific issues—and specific space-related experiment know-how.

Decision Phase

Based on the experimental idea and the resulting project plan, the company must decide for or against executing the project. This decision involves weighing the project against other alternatives and estimating its economic profitability. To achieve the same commercial objectives, ground-based alternatives to the microgravity project will usually be available. Ground- and space-based alternatives must be compared with respect to cost, period of performance, personnel commitment, and consequences for the company's strategic plan.

In considering the question of personnel commitment, it is important to consider that converting an experiment to a space-qualified version and conducting the microgravity experiment itself require considerable space experience. From the

---

*This inclination toward participation in space as a high-technology venture, coupled with the expectation of various spin-off activities, is much greater in the United States than in Europe.

authors' conversations with private companies, most users of space are not interested in developing this space-related know-how within the firm and would welcome cooperation with organizations that offer special space-related services. Exceptions include 3M, Wyle, Teledyne Brown, Grumman, Boeing, MBB-ERNO, Dornier, and IHI in Japan, which have space technology training programs in systems aspects (spatial and logical integration of equipment within limited resources), and mission assurance (including complete waste management).

A company's strategic considerations must include the question of availability of the necessary flight opportunities. It marks a major difference between microgravity projects and other R&D projects: to the already high risk of success in a microgravity project must be added the lack of control of physical access to technical infrastructure. Also, as noted before, pressure for success of a microgravity venture is greater than for ground-based solutions because of the undue publicity and high visibility of a space enterprise. Different companies have different attitudes toward public visibility.

In the cost comparisons, the value of government initiatives (discussed in detail in Chapter 11) are also included. These initiatives could be limited to providing a flight opportunity and the use of the corresponding infrastructure, or providing support for R&D activities within the company. The high costs of flight opportunities, in particular, will usually favor ground-based alternatives, at least initially.

During the authors' surveys in Europe, the United States, and Japan, no companies considered cost to be the main reason for their reservations about microgravity projects. Much more important was the inability of a company to foresee the benefits it might accrue from microgravity research. In case of a commercially interesting idea, funding itself was a secondary problem, although government financial and technical support will certainly have a motivating effect. This clearly suggests that the bottleneck in commercial utilization of microgravity is still in the first phase of a project—finding a suitable idea to pursue.

Experimental Planning

In this phase, intensive interaction begins with the external providers. Suitable flight opportunities have to be negotiated with national or international space agencies based on the nature of the mission and the projected date of the mission.

If multipurpose facilities cannot be used without modification, the experimental facility has to be designed and built in cooperation with the space industry. Together with other advisory and support organizations, details of the experiment must be planned, technical problems identified, and appropriate solutions developed. This requires effective technology transfer: the knowledge and experience gained in all previously conducted microgravity experiments must be available to the company. Such advisory and support institutions include

- Centers of Excellence and CCDSs (in the United States)

- Experienced investigators from universities and other private companies

- User centers established by the official space agencies as part of the infrastructure, such as MUSC and MARS (in Europe)

- Applications laboratories of the space companies

- National and international centers responsible for mission planning and operation, such as mission control centers.

Technology transfer can also be effectively achieved through databases of microgravity experiment technology.

The necessity for intensive information exchange has to be balanced by the proprietary rights of the company over the objectives and technical details of the experiment. Such protection of industrial secrets is essential to commercial space utilization. How well suited the above institutions are to supporting commercial microgravity projects will depend on their ability to meet both these opposing demands. Such clear regulations or even model contracts between these institutions and industrial customers currently exist only within the CCDSs in the United States [104].

The results of the planning phase will yield a project organization plan and an interface agreement that clearly specifies the interaction between the project partners (private companies, space agencies, space industries, financial and technical support institutions) in scheduling, financial responsibilities, and deliverable results.

From the above description, a very complex scenario emerges for developing a microgravity project. Here private companies, such as INTOSPACE (in Germany) and Payload Systems, Inc., and Instrumentation Technology Associates (in the United States) can assist a company by providing a tailored package of external services.

Ground-Based Preparative Activities and Accompanying Research and Development

Microgravity experiments are most significant if they are part of a larger systematic program to study the effect of gravity on the behavior of a specific material or process. The few microgravity experiments presently feasible, given the limited number of flight opportunities, in some respects represent only the tip of the iceberg, as suggested by the receipt of over 200 proposals for the German Spacelab-D2 mission, of which only about 70 can be accommodated. Accompanying ground-based experiments will help integrate the microgravity activities within the regular R&D activities of the company and create a reference base for comparison with the microgravity experiment results.

These ground-based experiments also include experiments in drop towers and on research aircraft (parabolic flights), experiments in centrifuges (higher rather than reduced gravity), and/or experiments in the containerless processing of liquids that can be levitated by electrical, magnetic, or acoustic forces. Theoretical studies, computer simulation, and modeling studies will also contribute to understanding gravity-dependent phenomena.

Converting the Laboratory Experiment to a Space-Qualified Experiment

The space-related experimental technological problems are solved in cooperation with the space agency providing the flight opportunity, the hardware manufacturer, and the institutions named above under "Experimental Planning." Such problems might include mounting the sample assembly in the experimental facility, meeting safety requirements, testing stability under the vibration loads that occur during launch, and preparing operating procedures and training astronauts. These activities have very little to do with the actual scientific objective of the experiment, but they do require considerable space-related expertise. As noted earlier, the private companies generally prefer to obtain external support.

During this phase of the microgravity project, detailed operating procedures for the astronauts and/or computer programs for automatic control of equipment are developed and tested. For these tests, functional duplicates of flight equipment must be available on a timely basis and operated by trained personnel.

To perform microgravity experiments, a wide spectrum of suitable low-gravity facilities must be available. Companies must evaluate the available flight facilities in light of frequency of reflights, flight duration, achievable microgravity quality, cost per experiment and flight, and manned vs. automated flights. Most important for commercial activities are frequent, regular reflights, a reliable launch schedule, and the shortest possible times between sample delivery and flight and between flight and sample return. The launch schedule and the experimental facility must be optimized: for example, simultaneous processing of several identical samples for statistical reliability, or processing larger quantities of materials. The operation of the space hardware and the corresponding ground-based apparatus should be similar to ensure the utility of microgravity results for more advanced developments of the process and materials on Earth.

The industrial investigator may also wish to follow the performance of the experiment during the flight from the mission control center, and, if necessary, to make changes during its operation. This requires that the corresponding facilities (data acquisition, telemetry links, and so forth) are available in the respective control centers. A major requirement is to organize the data flow during the experiment to maintain the proprietary interests of the company.

Here also the customary uncertainties associated with ground control of the mission also conflict with industry's desire to estimate as closely as possible all costs and risks of a project. This will be briefly described using the Spacelab-D1 mission as an example. Highly automated missions, including TEXUS, reduce these uncertainties.

As a prerequisite to entry to the mission control center, the user has to participate, as a rule, in one or two simulated mission training runs, each of which lasts two to three days with two 12-hour shifts per experiment per day. Whereas schedule can be controlled somewhat reliably in these simulated runs, the actual mission may be delayed by several days or even several weeks. This can mean several trips to the control center, or in the case of particularly sensitive samples that require the investigator's presence at the launch site, a longer than planned stay in the United States. Even after launch, exactly when the experiment will be conducted can be uncertain, perhaps requiring a permanent presence (two shifts) for seven to ten days in the control center.

This boundary condition must be considerably simplified for future missions, especially for future operation of Space Station Freedom. One method could be to control the experiment directly from the investigator's own laboratory by telemetry.

Performance of the Microgravity Experiment

The results are evaluated in the company laboratory using the space samples and the experimental and spacecraft data. It is essential that the sample be returned expeditiously along with all recorded data. Here also uncertainties inherent in most current missions make it difficult to estimate the effort required. For comparison with the microgravity experiment, a reference experiment using the flight version of the experimental apparatus must be conducted on Earth under real-time boundary conditions. Here again support will be needed from space agencies, user

Evaluation of the Results

centers, or the space industry to operate the apparatus on Earth, as in the conversion phase (described previously).

Exploitation of the Results

The results of space experiments conducted by academic institutions can be published in the technical literature and/or in the popular media. Industrial companies are more interested in using the results internally for material and process improvements and, combined with patent filing, to give the company a technological advantage over its competitors. At this point a conflict of interest arises between the government and the private sector: although private industry wants to retain the data for its exclusive use for as long as possible, the government is interested in the widest possible dissemination of the results to benefit the most users and to gain new users for space infrastructure. If a company pays all costs, including its share of mission costs, it alone has the right to decide about the disposition of data. This will seldom be the case, at least until space operations become inexpensive routine ventures. In most cases, the government, using NASA's Joint Endeavor Agreements (JEA) as a model, will provide the mission opportunity and will share in providing the experimental facilities, particularly multipurpose facilities. Thus, the government share of space experiments will always be in excess of 50 percent, on an average as high as 80 percent. How this will affect commercial utilization of the results is still unclear, since no single German or other European private company has as yet participated on a sizeable scale in microgravity missions. NASA provides JEA customers with very generous terms and conditions (see Chapter 12). One possible compromise between government and industry could be limited-term copyright, the length of the term dependent on the company's financial contribution to the project.

REQUIREMENTS FOR UTILIZATION

From the above description of activities in the individual phases of a microgravity project, it is possible to derive the requirements of a company to profitably utilize microgravity. These requirements are briefly described in the following sections.

Market

Commercial utilization of microgravity is only feasible if markets exist for the microgravity products, or if other terrestrially manufactured products can achieve a market advantage by improvements derived from microgravity research. To make these decisions, the private company must be able to project, quantitatively and in advance, the market size for space products or the market advantages over a longer time.

Possibility for Progressive Entry

In view of the current difficulty in assessing the market situation, private companies must have the opportunity (for example, through small, less costly missions) to become involved in microgravity utilization gradually and progressively.

Accountability of Project Performance

Project performance must be amenable to review and cost accountability so that development costs can be compared with projected market size. This requires long-term guaranteed access to flight opportunities and an accountable, long-term pricing policy by the providers of flight opportunities.

96

The scenarios covering the interactions of the company with other organizations must be simple. The administrative effort involved must be kept to a minimum for voting decisions and for negotiations about the nature, extent, and schedule to meet significant milestones. Review of proposals for experiments, negotiations of flight opportunities, and eventual financial assistance must be completed in a timely manner.

Management of Scenarios for Interactions

The company must be able to acquire the know-how available at other organizations, including the physico-chemical effects of microgravity and technology for experimentation under microgravity.

Technology Transfer

The company may require external technical and administrative support, particularly to convert the laboratory version of the experiment to a space-qualified experiment, to integrate the experiment into the total space flight system, and to perform the experiment. This support includes actual performance of the work by trained specialists and special equipment and facilities.

External Support

A wide variety of suitable flight opportunities (see Chapter 13 under "Experimental Options for Research in Microgravity") must be available. A commercial microgravity project generally cannot be accomplished in a single space experiment, but rather will be a comprehensive program involving several flights and extensive ground-based work. For example, critical components or sections of an experiment could be tested in parabolic flight and preliminary experiments with the entire experimental assembly can take place on sounding rockets (such as TEXUS) before the actual experiment is flown on a large mission like Spacelab or EURECA. Even for a single project, a variety of flight opportunities are required. Also, the space experiment will be repeated several times to obtain statistically significant data or to produce sufficient quantities of material.

Flight Opportunities

The nature and methodology of the experiment must correspond to the usual procedures for performing terrestrial R&D projects in the industry. In contrast to basic academic research, an industrial user has little or no leeway in adapting experimental parameters (for example, the material to be studied, the duration of the process, or the choice of experimental apparatus) to the technical boundary conditions of existing facilities. Rather, these boundary conditions must be adapted to fit the design of the industrial experiment. The margin for compromise is smaller here than in basic research.

Industry-Oriented Experiment Strategies

After the mission, the space samples must be returned to the user company quickly. All data generated during the mission must be provided to the investigator in "user friendly" form.

Post-Flight Activities

All data related to the project must be treated confidentially.

Protection of Company Data

Utilization Rights The rights to use the results belong entirely to the investigator.

Financing The uncertainty of commercial utilization combined with the high cost of microgravity projects represent a large risk for a private company. At least in the initial phase of commercial utilization, companies that have pioneered new achievements by early involvement in microgravity rely on the government to cover a part of the risk through

- Assumption of individual cost elements in the form of subsidies. This generally covers flight costs, integration of the experiment in the flight system, and elements of the support infrastructure.

- Guarantee of credit to be repaid in full or in part after commercial success of the venture.

Even in the later phases of commercial utilization, the financial risks of a microgravity venture will be much higher than comparable R&D projects on Earth. The company must be able to obtain the capital to cover these risks on the capital markets. Government guarantees can also smooth the path to microgravity utilization for small and innovative, but financially weak companies.

CONDITIONS NEEDED TO MEET REQUIREMENTS

These requirements can be met only if specific political, legal, economic, organizational, and institutional conditions are present, as represented by the existing infrastructure.

Market growth depends very strongly on how the cost issue is perceived by the user company. Cost reductions are effected through

- Development of reasonably priced space transportation systems.

- Provision of orbital missions at reasonable costs. This can be achieved by tailoring the missions to the individual projects or a typical class of project.

- Establishment of an effective operating organizational infrastructure.

To make it possible to **evaluate project progress,** the infrastructure elements discussed below must not only be available and effectively organized, but the infrastructure network must also be visible to the user.

A clear **conceptual framework** must exist, within which the individual infrastructure elements can be organized, to regulate the mutual responsibilities of the participating institutions. Examples of successful commercial microgravity projects should be available as models; these will assume prior experience and familiarity of the participants with microgravity projects. Experience with academic microgravity research projects is applicable only on a limited basis to develop such models.

Guaranteed access to space should be viewed as a political and technical issue and resolved as such. A national policy for access to an STS, to infrastructures in space, and to ground operation and control systems is essential. Such a national policy and operating system can adequately substitute for an independent indigenous system through recognizable, long-term, stable international cooperation, such as the planned Ariane, Hermes, and Columbus programs within ESA. What is important is that the user understands this policy. Such international cooperation with the same partners must also have been proven in other areas. National space policy must show continuity and consistency.

Access to space must also be technically assured: technically reliable space systems are required. The loss of Challenger and the subsequent unavailability of Ariane, which comprised nearly the entire European and U.S. launch capacity during the hiatus, considerably dampened the enthusiasm of many potential space users. (The consequences of the Challenger accident were discussed in Chapter 9.)

To keep the multiple **interaction scenario** operating effectively, the programs and operations of the participants must be harmonized. This is particularly true for international projects, such as programs utilizing ESA missions for nationally supported projects.

General support programs must be adapted to, or special support programs developed for, the atypical management scenarios and conditions of microgravity projects.

For the user company to organize its space R&D activities from a broad base of **fundamental knowledge,** extensive basic research on the effects of microgravity is necessary. This fundamental knowledge is not necessarily obtained exclusively from microgravity experiments performed by academic researchers. More important to the user are "practical" investigations of the effects of gravity, such as convective flows and hydrostatic forces in fluids, through a variety of experiments on Earth and from theoretical studies. This part of basic research is particularly amenable to integrating microgravity research into the "normal" ground-based research, thereby supporting the commercial user in

- Generating ideas for microgravity experiments based on technical and scientific problems from the company's general terrestrial R&D activities

- Applying the knowledge from microgravity experiments to improve terrestrial manufacturing processes.

The know-how obtained from basic research and from other commercial projects must not only be available, it must also be processed and be in usable form. As far as knowledge of the effects of microgravity is concerned, this can be in the form of accessible technical publications and reports and briefings by the scientists involved. However, to perform a microgravity project in a cost-effective manner, the commercial user must be able to learn quickly about the newest developments in weightlessness technology. Generally, the user will not find sufficient practical information in scientific publications. What is more useful are special databases or handbooks on microgravity experiment technology and close cooperation with experienced investigators from the academic community to support direct information exchange.

To provide **administrative support,** and above all **technical support,** to the user, government or even private organizations must provide expert advice about possible flight opportunities and their specifications and operating conditions relevant to the project. Converting the laboratory version to a space-qualified experiment requires considerable specific expertise in space, which the user company cannot develop at the required level (it is cost-effective to use an external service). Therefore, government or private organizations must also help the user to

- Design and prepare material samples, measuring cells, and components for the experiments.

- Modify existing apparatus for the proposed experiment.

- Perform the required tests (for example, test runs of the experiment) to define the process parameters and to optimize the sample assembly. Also perform qualification tests to demonstrate the flightworthiness of the experiment.

- Integrate the experiment in the flight equipment and integrate planned experimental operations into the mission plan.

- Provide the very extensive documentation that will be needed, such as operating procedures for the astronauts, manufacturing certificates for the experimental components, manufacturing records, construction records, safety analysis, and test certificates and reports.

- Train the astronauts specifically for each experiment.

The facilities and equipment required for these activities include functional duplicates of the flight apparatus to develop and test the experiments and special test facilities (for example, for vibration tests).

To perform these activities cost effectively, expertise in three areas must be combined:

- Knowledge of the goals and objectives of the proposed project within its commercial framework

- Technical knowledge of the flight equipment of the mission and the experimental apparatus

- An understanding of the experiment technology and of how to use it to develop the corresponding project.

The first item must be covered by the investigator. Information related to the second point can be provided by the space industry. The support organizations discussed above cover the information related to the third item. These organizations link the user, mission providers, and the space industry. Their function is not only to perform the above mentioned tasks, but also to manage and coordinate the activities of the other participants in the project. To cover the third item, the user should consider experienced experimenters from universities and the scientific community, private or government user centers that specialize in these tasks, and/or applications laboratories of the space industry.

Flight opportunities are distinguished by the following parameters:

- Flight duration

- Frequency of reflights

- Achievable microgravity quality (g-levels)

- Manned vs. automated operation

- Available capacity in terms of size, weight, power, and data acquisition and handling per experiment

- Flight cost per experiment or per kilogram of payload.

To provide the user with a flight opportunity tailored to the experiment at reasonable cost, a wide spectrum of flight facilities with different values of the above parameters is needed. In general, a commercial user needs high reflight frequencies to avoid long waits between flights, as well as to perform series of experiments, a regular flight schedule (as far as possible), and a highly reliable flight schedule.

To prepare for complex experimental programs, in particular, small inexpensive suborbital missions with very high reflight frequencies will be required, primarily parabolic flights in research aircraft. These suborbital missions can be used to develop and test individual subsystems for the experiment, thereby reducing the failure risk on the subsequent major orbital flight missions.

These suborbital missions, as well as ground-based drop tube and drop tower experiments and simulation of weightlessness by containerless processing (positioning of samples by magnetic, electric, or acoustic forces), despite their poor experimental boundary conditions, provide an excellent opportunity for the commercial user to ease into microgravity research.

The necessity of adapting the experimental apparatus and the operating conditions to fit the commercial interests of the respective experiment—a greater need in microgravity than in basic research experiments—requires a high degree of flexibility on the part of the mission provider and the space industry. Thus, it is important to move away from multipurpose, multi-user facilities toward development of individually tailored facilities [112].

The choice of materials for processing in space must be based on economic criteria. This means that even hazardous materials cannot be excluded from space missions. Suitable safety plans and technically safe facilities are required. The safety problem may become even more acute if the company cannot provide material safety data for proprietary reasons (explained below).

The demand for rapid return of the space samples can be fulfilled by mission managers for most missions. In the case of a permanently operating space station, however, frequent return (every three months) of samples must be possible.

For **processing "user friendly" data,** the corresponding facilities for data acquisition and processing must be available in the mission control center. During the mission, the user must have on-line access to the data and react rapidly, perhaps by remote control or through contact with the astronauts who can adjust the experiment. Data handling (acquisition, preparation, and processing) must be organized to protect the proprietary interests of the user.

To **protect proprietary interests** of the user, practical models must be created that reflect basic user interests and the mission manager's need for safety-related information. Here not only must legal conditions be developed, but technical issues related to the type and extent of the qualification program,* basic safety philosophy, and the technical facilities to guarantee flight safety must also be considered. For example, if the company provides all material safety data for the experiment, it is usually possible to estimate the safety risks of toxicity or flammability from reference data. On the other hand, by keeping the data secret, the safety information must be experimentally determined, or expensive prophylactic technical safety features must be added.

The **legal problems** related to guarantee of proprietary rights, mainly patent and other intellectual property rights, are comparatively easy to solve with

---

*The qualification program proves the flightworthiness, particularly the mission safety, of the experiment. This program is carried out in the project phase to convert laboratory experiments into space-qualified experiments.

missions managed and controlled by the user country with nationally controlled STS. For international missions and particularly for the proposed Space Station Freedom, clear regulations are indispensable and corresponding agreements must be reached between the participating nations. These issues are considered briefly in Chapter 12.

To cover the financial risks, unambiguous **financing mechanisms** are essential. The government can use these mechanisms to provide the necessary financial resources, initially in large amounts, then decreasing progressively with increasing degree of commercialization. Financial subsidies are needed mainly to cover those specific space-related costs that make microgravity research more expensive and less competitive than terrestrial research. Besides flight and infrastructure costs, these include the company's internal costs associated with project preparation and performance.

Direct subsidies for ground-based development are required at the same level as funding guaranteed for comparable R&D projects on Earth. However, the conditions (for example, project duration, thematic emphasis) of the existing R&D support programs are often incompatible with the specific boundary conditions of microgravity projects [14]. Such compatibility is a basic precondition for the usefulness of these support programs for microgravity projects.

CHAPTER

11

Economic Framework Conditions

A general discussion of economic framework conditions was presented in Chapter 4 for satellite remote sensing. In this chapter, only those aspects specifically relevant to microgravity utilization will be discussed.

Microgravity experiments (with the possible exception of some fluid physics experiments) are distinguished from other space utilization disciplines by the necessity of returning the samples to Earth and by repeated use of instruments after servicing them on Earth [115]. For an experiment in microgravity, four groups of costs must be considered:

COST AND FINANCING

Group A. Cost of developing the space technology.

Group B. Cost of constructing and operating the space flight system, including

- Construction of the launch system facilities and the payload elements

- Flight costs for the STS

- Mission control

- Operation and control of the experiments from Earth

- Data acquisition

- Management.

Group C. Cost of integrating the experiments in the space flight system, including

- Converting the laboratory version to a space-qualified experiment (described in the previous chapter)

- Integrating the experimental hardware (experiment module and payload element) into the flight system, acquiring the necessary support facilities, and conducting the required tests.

Group D. Cost of developing a laboratory experiment and the accompanying ground-based research.

These costs will vary widely with the specific requirements of each experiment. The costs of microgravity in terms of the launch costs per kilogram or per kilogram-hour are shown in Figure 8. Costs range from U.S. $1,000 per kilogram of payload for the TEXUS sounding rocket (the experimental microgravity time is minutes) to an estimated U.S. $4,000 per kilogram of payload with free-flying platforms like EURECA, where the experimental time can be up to six months.

By presenting the costs as shown in Figure 8, given the presently available experimental times per kilogram-hour, the need for Space Station Freedom as a low-cost facility becomes very clear. The costs become even lower by using unmanned free flyers. Similarly, g-levels (residual gravity) and cost per kilogram-hour also decrease with flight facility. But this desirable relationship will have an impact only if the long experimental microgravity durations on S.S. Freedom and the free flyers are actually engaged by the users.

Based on another study [92], the estimated costs for space flight, mission control, and payload integration per experiment are given below for various European flight opportunities.

Flight opportunity	Microgravity duration	Cost (in thousand DM)
Parabolic flight (preliminary experiments)	seconds	200
Sounding rockets (TEXUS)	minutes	350
Space shuttle (GAS/MAUS)	hours/days	450
Space shuttle (SPACELAB)	days	1,800
Free flyer (EURECA)	6 months	5,000

Not included in these figures are costs in Group D for preparation of the experiment, development costs (Group A), or costs of the infrastructure.

These costs are also valid on an international basis in their order of magnitude. However, these costs may change as new and more cost-effective launch options become available in the future [17,18].

A comparison of the two relationships in Figure 8 with the information in the above table indicates that an optimum price level depends on providing the user specific flight opportunities designed to meet individual user needs. For example, sounding rockets or parabolic aircraft flights would be too expensive for manufacturing projects where larger quantities of materials must be processed for long times in high-quality microgravity. But for a single experiment, sounding rockets, aircraft parabolic flights, and Shuttle payloads in GAS cans are a bargain. These flight opportunities can be preferentially utilized in the starting phase of an experimental series, when the quantity of material for space processing is still small.

Ideally, it is possible to speak of commercial utilization only if the user covers all four of the above cost areas from company resources. This ideal situation presently does not exist in any country and probably will not exist in the near future (compare also Chapter 9, the section entitled "Demand for Utilization of Microgravity Capabilities"). However, different countries give the initial indications of at least partial commercialization, as companies are investing their own resources in the individual cost groups.

According to the general BMFT funding policies in Germany, private companies who have experiments on the Spacelab-D2 mission are assuming 50 percent of the accrued costs for the experiments. A good example is the joint project, OSIRIS, between industry and academia.

As noted earlier, Japanese firms participating in the German Spacelab-D2 mission have formed the Space Technology Corporation (STC) and provided 30 percent of the starting capital.

In the United States, NASA has signed JEAs with several companies [104]. The activities planned by these firms (some activities have already been performed) are:

- Utilization of microgravity conditions for experiments and contributions to costs in Group D.

- Construction and marketing of experiment hardware, thereby assuming costs in Groups B and C. These companies are also

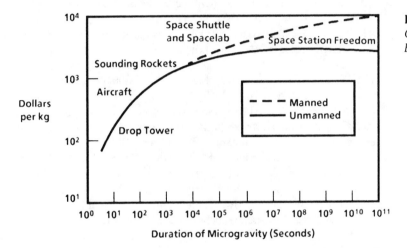

FIGURE 8.

Costs for Microgravity Experiments

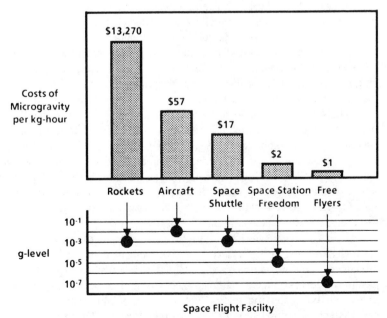

Source: R. Boudreault, Canadian Astronautics Ltd, 1985

105

assuming part of the risk of whether there will be sufficient users for the hardware; the McDonnell-Douglas electrophoresis project is a good example of a private company assuming the former risks. In other instances, this risk is assumed by the government.

In France (compare the discussion in Chapter 9), a group of companies is constructing a facility to grow protein crystals in space using its own financial resources [94].

In all of the countries considered, the government is basically willing to assume those costs that are specifically space-related and that make microgravity projects more expensive than ground-based R&D projects. They have, at least, covered the risk of commercial failure of the project.

Cooperative ventures between the U.S. government (through NASA) and commercial users are regulated by the JEA. Under a JEA, the private company pays the cost of salaries and the experimental equipment, and NASA pays the cost of launching the initial experiments into space aboard the Shuttle. The JEA allows the company access to NASA equipment and facilities, and the firm gets a specified number of free flights. NASA's guidelines for JEAs were originally published in 1979 [116]. Since the first JEA with McDonnell-Douglas in January 1980, seven other JEAs were signed under the original guidelines [117]. Recently, NASA has developed new guidelines for JEAs [118] under which the company can pay NASA for flights after it begins earning revenues from its microgravity activities. As of 1988, NASA had eight active JEAs with various companies [104].

Through this concept of "Fly now, pay later," NASA is already making efforts to commercialize space utilization in the broadest sense. At the same time, however, NASA is bearing a very large part of the still very high financial risk of a microgravity endeavor by waiving repayment in case of commercial failure.

A similar but less well defined cooperative effort can be found in Japan. The Key Technology Center (Key-TEC) guarantees financial credit to industry for research projects in key technologies and for projects from applied research to commercial maturity [119]. This financial credit can cover up to 70 percent of project costs. If the project does not provide any commercially useful results, no interest payments are due on the credit, as shown in Figure 9.

To what extent and which model the Japanese government uses to assume a part of costs (in the form of subsidies) of industrial microgravity experiments is currently known only for Japanese participation in the German Spacelab-D2 mission. However, it can be assumed that participation in other international and domestic missions is similarly handled. As noted previously, the companies participating in the Spacelab-D2 mission have formed a subsidiary, Space Technology Corporation (STC). Key-TEC owns 70 percent of STC [119].

The companies prepare their experiments within STC and also purchase flight opportunities and integration support on completely commercial terms. To what extent the 30 percent industrial share of STC also covers the company's costs for preparation of the experiment, or if the company does provide additional resources, could not be determined.

In Germany, even industrial experiments, such as those on the Spacelab-D1 mission or on TEXUS, have been 100 percent financed from government resources. However, as noted earlier, BMFT is planning to require that industry assume at least 50 percent of R&D project costs. This means that costs in Group D will be borne by the microgravity user, although under the appropriate circumstances "normal support" will be guaranteed by the government, as with terrestrial R&D projects. Costs in Groups A and B are assumed by the government.

In this respect, Germany is not very different from its ESA partners. Even in France, the experiments are financed 100 percent by the government through CNES. The same is true for Italy. The single experiment of a British firm (Kodak, Spacelab-1 experiment) was financed 55 percent by Kodak and 45 percent by government grants [120]. France, Italy, Great Britain, and Japan have conducted so few experiments—compared with the United States and Germany—that predominant cooperative financial models are not apparent.

The situation for costs in Group C is less clear in all countries, since this cost group includes costs for the infrastructure that the user must employ to convert a laboratory experiment to a space-qualified experiment.

The costs for developing, and usually also for operating, the infrastructure are still borne by the government in all countries. User costs depend very much on the quality of this infrastructure. Thus, the government can provide indirect financial support to the user by investing in additional infrastructure. Different countries handle this phase in distinctly different ways, which will be discussed in Chapter 13.

When a company weighs a microgravity R&D project against a possible terrestrial alternative, the framework condition costs are the additional costs associated with the use of microgravity. In this respect, any differences in the R&D support programs, which also apply to other R&D projects, do not represent any differences in the framework conditions for utilization of microgravity. As was repeatedly stated during the authors' discussions with private industry, increased

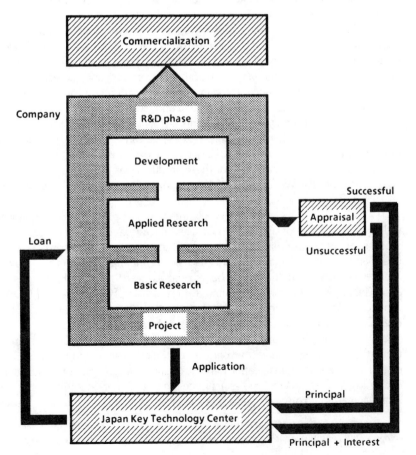

FIGURE 9.

Typical Flow Diagram for a Project Supported by the Key Technology Center in Japan

financial support alone would in no way stimulate commercial microgravity ventures. Pressure for commercialization through use of financial incentives would significantly hinder market-oriented selection of industrial microgravity projects and would not lead to self-sustaining private sector involvement in microgravity. However, it should be noted that microgravity projects, for example, because of their long duration, often do not fit easily within the normal R&D support programs [14]. Therefore, differences in framework conditions can exist, if the existing support programs in the individual countries are well adapted to microgravity project constraints.

MARKET CONDITIONS

Unlike satellite communications and even satellite remote sensing, there is no existing market for microgravity. The research is still too young. To date, only one product (latex spheres) manufactured in the microgravity conditions of space is being sold commercially [86].

Even though no one assumes that this situation will change in the near future, all future projections for long-term development assume that commercial markets for microgravity products will develop from the results of further research in microgravity, and increasing demand for advanced products [9,86]. The activities in which markets for space-processed materials can be expected to develop have been mentioned previously in Chapter 9 in the discussion of the possibilities for utilizing microgravity. These activities include products manufactured in space that can be sold directly on the market and R&D in microgravity that can improve the market potential of products manufactured on Earth.

With respect to manufacturing in space, only those applications in which there are clearly defined products for large existing markets have a chance to develop relatively quickly. Pharmaceuticals and electronics are prime candidates [9]. Product ideas for pharmaceuticals will result primarily from cell and biotechnological production in space, mainly through improved separation of components (microgravity conditions enhance selectivity and yields).

A market analysis by McDonnell-Douglas Astronautics Company (MDAC) identified 12 pharmaceutical products that can be profitably produced by electrophoresis under microgravity [121]. Even under conservative assumptions, the annual U.S. domestic market for these products was estimated in excess of $5 billion (Figure 10). World markets for these products are estimated at $23 billion by 1991 [122].

However, it is exactly in this field that a major problem is clearly evident with private sector involvement: in most cases, the absolute limits of the terrestrial process cannot be determined, making a direct cost/benefit comparison of ground-based and space production difficult. Also, significant improvements at far lower cost can be expected in the ground-based processes, so that the industry feels no strong pressure to be involved in microgravity processing. For example, the MDAC continuous flow electrophoresis system for cell separation in microgravity has been overtaken by alternative technologies such as chromatography [123].

This argument appears to apply also to the other most obvious single market for production under microgravity conditions, the manufacture of high-quality (lowest possible dislocation density) gallium arsenide* and other semiconductor single crystals. U.S. studies estimate the annual market for these crystals is growing at approximately 30 to 50 percent per year and is expected to reach

---

*Gallium arsenide (GaAs) is the basic material used for high-speed integrated electronic components for microwave and radar applications and for very high-speed data processing systems.

$1.8 billion by 1992 [1,9]. For gallium arsenide, potential market demand could be nearly $6 billion in the year 2000, of which nearly half may be military purchases. Here, also, the long duration of microgravity projects could make space production less promising, especially if successful advances are made in ground-based processing technology in the meantime. In fact, Microgravity Research Associates wanted to manufacture GaAs crystals in space, but has found ways to improve earth-based processing of high-purity crystals [124].

Both of these product groups possess an additional important characteristic of potential products for commercial space production: high price/weight ratio. This characteristic also has led recently to increased interest in the use of catalysts to accelerate or increase the yields from chemical processes. The high cost for producing catalysts such as zeolites in space are justified through the cost savings their use allows in chemical production on Earth [125].

Improvement of the market potential of products manufactured on Earth and the development of new products based on knowledge gained from microgravity research will precede actual manufacturing in space. This path to microgravity utilization will probably represent the major thrust of commercial microgravity activities in the medium term (to 2010). However, the potential market demand is difficult to estimate quantitatively.

R&D in space also includes medical research in the form of medical and psychological experiments on the astronauts or on test animals. The results of this research will lead to the development of new commercial products or new cures on Earth, as well as providing information to support people in space. The development of life support systems and medical care for astronauts can evolve into a limited market in a relatively short time. The size of this market will increase with the increased number and length of stay of people in space.

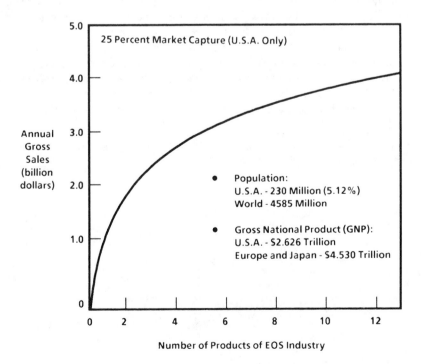

FIGURE 10.
Annual Market Potential for Electrophoresis Products

Source: McDonnell-Douglas

109

In summary, the following picture emerges for commercial markets for microgravity activities:

- Even in the short term (to 2000), a very limited size market for the results of medical research can develop to meet the demand for life support systems and medical care for astronauts.

- In the medium term (to 2010), the use of microgravity research results to improve products and processes on Earth will be the major thrust of commercial activities.

- Markets for space-manufactured products are expected to develop in the long term (beyond 2010).

Quantitative estimates of the development of these markets have been given by some companies, primarily firms who are also suppliers of space systems. However, all the quantitative projections are very uncertain. Hence, these estimates cannot serve as a reliable basis for potential microgravity users to make their decision for or against involvement in microgravity.

As critical as the development of markets for microgravity products and results is to commercial utilization of microgravity, individual differences in the prognoses for each country are minimal. The industries in all the countries considered routinely develop and produce technologically advanced products. The one important requirement for the creation of a market for microgravity—the demand for high-technology products—is present in all of these countries.

Regarding the market for products and methods for life support systems for astronauts, all the countries have made the decision to develop, operate, and use manned space systems. Even so, the potential demand for the domestic U.S. market probably is greater than for the European or Japanese markets. Until the Challenger accident, the United States employed a manned STS even for launching satellites with the Shuttle, but as of 1988, commercial satellites will no longer be launched on the Shuttle. The planned European transportation systems will be manned only if astronauts are to be transported to or from Space Station Freedom. Furthermore, the United States is also seriously considering manned missions to Mars [19,126]. Such a Mars mission will place extreme demands on a life support system, which in turn, will have a positive effect on market development.*

---

* That U.S. organizations cover material science tasks—whereas Europeans cover life science and medical tasks—aboard Space Station Freedom speaks to their high assessment of future market applications for microgravity material sciences.

CHAPTER

12 Legal and Political Considerations

As long as space experiments are within the realm of basic research, the results of which are internationally accessible, no major legal problems occur. The situation changes once a private organization invests its own funds in the project and claims proprietary use of the data. The problem occurs on two different levels: the issue of mutual obligations of the company and the institutions (DFVLR/BMFT, ESA, NASA, NASDA) operating the mission or their respective sections of Space Station Freedom; and the issues involved in the legal agreements between the international partners in an orbital mission or in S.S. Freedom [127,128].

LEGAL PROBLEMS

International Legal Problems

Isolated legal problems have occurred and will occur if, for example, American launch vehicles are used for German missions. In the future, such problems will not be solved on a case-by-case basis, but by negotiations based only on fundamental considerations. However, if one considers the planned S.S. Freedom to represent a single unit in the sense of a single flight facility, then any legal discrimination based on material issues does not appear very meaningful.

Individually, the problems of interest are those that must be regulated between federal agencies involved in the construction and utilization of S.S. Freedom [127,128].

Between the alternatives, "government agreement" and "international treaty," are a number of important differences. What is important here is that concluding an international treaty requires time-consuming procedures, but if the regulations are sufficiently detailed, it will not require any modification of national laws. This would protect the treaty partners from any undue interference by changes in national policy decisions. At the same time, regulations can be made to deal with problems to be expected in space, regardless of the circumstances on Earth. At present, the United States, unlike other partners, appears to be striving for a government agreement.

Registration of Modules. It is possible to register the modules of S.S. Freedom with one central authority, but the modules will then have to be assigned to one nation; separate registration could create a basis for assigning separate sovereign rights.

Property Control and Sovereign Rights. The separate registration of the modules, according to prevailing interpretations, is still not sufficient to justify separate sovereign rights. Control of the modules must also be secured. However, if the technical control of the operation and resupply of S.S. Freedom is not within the country that has formal control of its overall direction, that country's ability to follow through on its operational responsibilities is jeopardized. If, on the other hand,

111

technical control is a delegated service in a contract between participating nations, the result could be what is commonly referred to as the "legal balkanization of the Space Station"; that is, the different components of S.S. Freedom will have separate sovereign rights [127,128]. To put S.S. Freedom under a uniform national law (probably U.S. law) would probably result in the least cost and fewest procedural disadvantages for the other participants.

Property Rights. Actual property rights are not well defined with respect to space. These rights need not necessarily be associated with discretionary authority or with legal liability responsibilities.

Trade Restrictions, Taxes, and Tariffs. Regulations to avoid time-consuming procedures in these areas appear to be less problematic, but they may require special attention in isolated cases, especially with respect to commercial applications. The concern here may be to prevent competitive advantages for U.S. commercial users.

Intellectual Property, Patents, and Other Protective Rights. The patent laws of individual countries cover the use of space, a space station, or internationally used ground facilities in different ways. For example, if space equipment under U.S. sovereignty is considered U.S. territory, the participating American firms could more easily sue their European and/or Japanese competitors for patent infringements. Regardless of the validity of the claim, there is a high risk of a lawsuit under the applicable U.S. legal system with its customary high monetary judgments.

Access to Transport Systems. This point applies mainly to the freedom of the United States and NASA to determine launch dates and Shuttle prices.
A long-term joint project like the planned S.S. Freedom demands very well-defined cooperation. Regulations governing the assignment of priorities in launch planning and the modality of pricing should be in keeping with such cooperation. To what extent NASA is willing to consider such regulations is difficult to estimate reliably until NASA's space station policies are defined.

Liability and Insurance. The problems of third party liability are limited to countries registering space objects, according to the liability convention of COPUOS. The extent to which the liable country will indemnify a commercial user in the event of negligence during the project depends on the countries' willingness to avoid any uncalculated risks in commercial utilization. If the countries are not willing, then compensation claims must be regulated by the existing provisions of sovereign rights. Large losses resulting from launch vehicle failures have made space insurance more problematic, as illustrated by the steep increase in premiums for satellite launches from approximately 5 percent initially to nearly 20 to 30 percent of the total insurance value [9,129,130].

Summary. Difficult legal problems remain to be solved in a very politically characterized field. The continuing dependence on the United States of the other countries and institutions interested in microgravity, until at least the year 2000, has confronted the official agencies in those countries with the decision either to provide unilateral support for the interests of their own industries in commercial utilization of space, or to seek equitable international cooperation as the basis for developing the potential of space even further. A decision for the first option would become apparent in negotiations of the financial and legal conditions for the participation of other countries in the S.S. Freedom project. This option carries a long-term risk for the United States of strong future competition in space commercialization, since it would spur other countries to increase their efforts toward independent solutions to space transportation problems (*e.g.*, Ariane/Hermes for ESA).

Recent discussions in Europe about using Chinese and Soviet space carriers have shed some light on the murkier legal situation in international cooperation with these countries, since the political framework for such agreements is still elementary.

Legal Problems Between Private Users and the Government

Chapter 10, in the section entitled "Exploitation of the Results," briefly mentioned the existing conflict of interest between partners in a mixed government and private sector venture related to utilization of the results. This will probably remain a major condition for private sector involvement in space activities in the foreseeable future. Legal problems between the official mission operator and the private user occur not only with respect to protection of data rights, but are also related to other areas:

- Guaranteed access to flight opportunities (this point must also be considered within the context of competition between several providers of such services)

- Use of government facilities for experiment preparation

- Information exchange between the government and the private user

- Mutual provision of flight components and documents

- Safety review of items provided by the private company

- Project management and adherence to schedules

- Government reservations regarding the goals of the industrial experiment (*e.g.*, exclusively peaceful uses, advertising).

In general, the government, with its wealth of experience in conducting space experiments, has an interest in ensuring that a private company employs all existing conditions in the form of knowledge and facilities (*e.g.*, equipment to simulate gravity effects, preliminary experiments) for the success of the experiment. It will also provide the company with all relevant information. The only country in which concrete legal agreements between a government space authority (NASA) and a private user have already been developed is the United States. For example, the Joint Endeavor Agreement (described in the previous chapter) between NASA and MDAC for electrophoretic purification of pharmaceuticals in space regulates NASA's contributions as well as the mutual claims of both partners with respect to the above items.

With regard to the data sensitivity issues related to access and use, NASA grants very generous rights to its industrial customers. Under the terms of the JEA, NASA will not grant a JEA to another company for the same process (which is narrowly defined) until the original agreement expires. This viewpoint is also reflected in the U.S. Commercial Space Policy, which states that under certain conditions NASA will purchase the products from such a space venture [131].

Several types of follow-on agreements between NASA and private U.S. companies are available in the meantime [104]; these tend to reduce the concessions made to the private companies. Whether and to what extent the JEA and follow-on agreements can be used as models for other Western countries remains to be seen. The present situation—companies in Germany and France for the first time are prepared to invest their own funds for space experiments—is the time to work out standard contracts or agreements.

POLITICAL STRATEGIES

The political considerations for satellite remote sensing in the latter part of Chapter 5 are also valid in part for microgravity utilization, especially the conclusion that a country without an independent space transportation system (such as Germany) must make concerted efforts to guarantee reliable access to space for its commercial users. Unilateral dependence on quotas, providers, or other countries should be avoided whenever possible. In contrast to remote sensing, microgravity utilization (characterized by experiments of much smaller scale) depends very strongly on the continuity of the experimental program and consequently on the availability of flight opportunities. Any interruption in the flight program, such as the recent hiatus in U.S. Shuttle operations on which several microgravity programs (e.g., Spacelab and MAUS) depend, probably has a greater impact on individual industrial users than on the largely academic user community.

The sensitivities of the program become evident given that material and/or process improvements for which microgravity is to be utilized are constantly occurring in terrestrial laboratories. Exclusion of gravity is one of several options, which will be used only if it is available in timely fashion. Otherwise other options will be substituted. This is the fundamental difference between microgravity and other space utilization fields: a microgravity program cannot be based on a single space flight, and process advances are constantly taking place on Earth, so a flight opportunity that comes too late loses its competitiveness with terrestrial alternatives.

To provide opportunities for access to space, the countries considered have adopted somewhat different strategies thus far. However, as noted earlier, it is becoming more evident that the European countries in particular are striving to develop technical alternatives to achieve more political independence from the United States.

- The **United States** has in the Space Shuttle a transportation system that allows high reflight frequency (9 to 12 flights per year are planned over the next ten years [17,27]), and provides an adequate capacity for transportation in both directions (launch and retrieval). The United States is thus politically independent, but the technical dependency is a disadvantage (see Chapter 9 on the consequences of the Challenger accident). Since 1988 the U.S. has used a mixed space transportation fleet of the Shuttle and commercial small- and large-capacity ELVs to meet future foreign and domestic launch needs [17].

- **France** leads development of an independent launch system (Ariane/Hermes) in Europe. It plays the same role in launch systems that Germany has assumed in microgravity. Furthermore, France has secured a relatively high degree of political independence in access to space carriers through bilateral cooperation with the United States, the Soviet Union, and more recently, China. As far as launch systems are concerned, however, the technical alternatives are not so distinct:

bilateral cooperation with the United States leads to the same system (Shuttle-based); the Soviet Union does provide launch options, but only limited capacity for retrieval; however, samples can be returned to Earth from the Soviet MIR station. Chinese launch options involve the use of unmanned retrievable capsules to return payloads to Earth.

- To date, **West Germany** has opted for the U.S. Space Shuttle system for its orbital flights for microgravity research. At the same time, Germany has also developed an excellent highly reliable system for suborbital missions in the TEXUS program. Here again the large programs are politically and technically dependent on the United States for orbital missions. During the Shuttle hiatus, Germany was the only Western space nation with an operational space transportation system for microgravity payloads (TEXUS). In the meantime, Sweden also developed an operational sounding rocket program (MASER).

  Within ESA, Germany is also participating in the Ariane and Hermes programs and has proposed SÄNGER as a logical follow-on manned system to Hermes [132]. As discussed in more detail in the next chapter, the possibility of bilateral cooperation with the Soviet Union and China is being considered. Other projects are also in preparation (partly in collaboration with Italy) to develop retrievable systems (Raumcourier, Revex) and alternative launch systems (Scout rockets, project TOPAS). From these activities, it is obvious that Germany has recognized the limitations of its unilateral political and technical dependence and, like France, is developing other alternatives for access to space.

- **Italy** and **Great Britain** are participating in the Ariane and Hermes projects through ESA. Britain has also proposed HOTOL as an alternative to Hermes [108]. Italy is working together with Germany on the TOPAS project (see Chapter 13 under "Experimental Options for Research in Microgravity"). Carina is being evaluated solely by Italy [22] as a reentry vehicle for microgravity projects.

- The **ESA** countries together are seeking more autonomy and other technical alternatives with Ariane and Hermes, as well as the contribution of the Columbus Attached Laboratory Space Station Freedom [66]. Ariane is an expendable launch system. Retrieval capacity is still to be developed, perhaps with the Hermes reusable spacecraft or with retrievable capsules such as Raumcourier, Revex, or similar systems. The individual elements of Columbus have been so conceived that

  —Ariane can be used as the carrier (particularly for the Columbus Free-Flying Laboratory).

  —Materials and people will be transported in both directions by Hermes.

  —Important elements of Columbus (the Free-Flying Laboratory and unmanned free-flying platforms) can be operated independently of S.S. Freedom.

- **Japan** must still utilize external launch services at the beginning of its microgravity activities, but is very successfully developing an independent launch system for satellites (H-II). This system in combination with other planned retrieval systems, such as HOPE [80], could also be used for microgravity.

Several recent publications have discussed the special features of microgravity utilization and strategies to involve industrial users. Some of these frequently mentioned strategies, which have been partially implemented in the leading microgravity countries (United States and Germany), are listed below [9,14,86,91,114,121,131,133,134,135].

- Establish industrial user centers to support development of suitable demonstration missions, joint projects between university scientists and industrial experts, the development of experimental hardware, and data collection and transfer. The United States has established the Centers for the Commercial Development of Space (CCDS); Germany, the Centers of Excellence; and Japan, the Japan Space Utilization Promotion Center (JSUP).

- Evaluate information about the possibilities and potential of microgravity and the experimental results obtained to date.

- Increase the number of low-cost flight opportunities to generate continuity.

- Invite industry scientists to perform joint demonstration projects.

The above strategies include measures to provide the appropriate infrastructure for utilizing microgravity. They are basically similar in all of these countries. Individual differences in the degree to which these strategies are used will be discussed in the next chapter on organizational and institutional infrastructure.

CHAPTER

13 Organizational and Institutional Infrastructure

Through their microgravity programs, each of the countries considered has acquired different levels of knowledge of the effects of microgravity and experimental technology. Differences also exist in the processing and transmission of this knowledge.

As mentioned earlier (Chapter 9), Germany has the most extensive base of knowledge of microgravity in Europe and is probably comparable with the United States. This knowledge was gained through two Spacelab missions, FSLP in 1983, the first Spacelab flight under ESA control, and D1 in 1985, the first mission under German control. In addition, a large number of experiments have been conducted in the unmanned TEXUS program and within the MAUS program in GAS cans.

Most of the scientists involved have been from universities and scientific institutions, but industrial researchers have also conducted microgravity experiments. As a result, technical advocates are present in individual firms who are familiar with the technical potential of microgravity and who can relate this potential to the commercial interests of their firms.

The detailed work of preparing the university experiments generally has been done by masters and doctoral candidates who now make up the main group of resident researchers with the expertise in microgravity experimental technology. This situation may help or hinder the transfer of essential microgravity commercialization technology to industry:

- If the experimenters join private industry after completion of their studies, they can, as technical advocates, create a favorable environment for evolution of ideas for microgravity research.

- Scientists departing industrial positions take part of the knowledge and expertise with them.

Two factors have increased the risk of the loss of knowledge and expertise:

- The continuity of the experimental series designed for the Spacelab and EURECA missions was interrupted when Shuttle operations were suspended. Doctoral students with experience from the FSLP and D1 missions were completing their studies before they could transfer their knowledge and experience to the D2 and EURECA experimenters.

- At present, no systematic collection and processing of knowledge is available. Experimental results and theoretical interpretations are published in the scientific literature, and information on experiment technology is published in reports to the sponsoring agencies (BMFT through DFVLR as project manager). However, no operating system exists to consolidate the widely dispersed knowledge and information

DEVELOPMENT OF BASIC TECHNOLOGY

organized according to technical areas, such as specimen design and operating parameters. This problem has been recognized: DFVLR, ESA, and INTOSPACE are looking into the possibility of developing databases to collect all available worldwide information on microgravity.

An additional possibility for technology transfer lies in direct cooperation between experienced experimenters and potential commercial users. BMFT is attempting to develop joint projects in which universities and private industry work together on topics selected by mutual agreement within a larger field. Efforts such as OSIRIS influence microgravity utilization in several ways:

- They bring together commercial problems defined by industry and solutions obtained through basic research that may lead to new, perhaps commercially oriented, ideas for microgravity research.

- Industry obtains direct access to available scientific, technical, and administrative know-how in microgravity utilization.

In the other European countries, the situation is qualitatively similar to Germany, but the knowledge base is quantitatively smaller and narrower in focus, corresponding to the different emphases in the national space policies of these countries. France is not too far behind Germany, even though independent microgravity programs are lacking. French experimenters have been intensively involved in ESA programs and, as mentioned earlier, in research on flight opportunities with NASA, the Soviet Union, and China. The number of experimenters in Great Britain is clearly smaller; Italy has a position somewhere between France and Great Britain.

The situation in Japan is completely different from that in Europe. Since Japan has conducted few independent microgravity experiments to date, it has developed little independent expertise in microgravity. Proposed Japanese experiments and the equipment for them (such as FMPT) clearly follow the European and American patterns of development.

Still, Japan is not simply repeating European and American experiments. Much more evident are intensive efforts to make up the European and American lead even before the launch of its first dedicated mission:

- Japanese experimenters from industry are establishing close coopera- tion with experienced scientists from Europe and the United States.

- Japanese organizations are analyzing existing resources for processing scientific and technological information in microgravity, such as data- bases [136].

- The Japanese are obtaining an overview of the theoretical models for microgravity and computer programs for mathematical simulation of microgravity effects, as well as conducting specific experiments developed in Europe and the United States.

- Japanese institutions are analyzing previous U.S. and European experiments in terms of technically critical points, approaches to their solution, and the success or failure in flight [101].

Similar systematic analyses are currently nonexistent in Europe: no overview of all previous experiments has been assembled, nor have they been analyzed with respect to consequences for reflights, though individual aspects have been considered separately [91].

The United States has a longer tradition of microgravity utilization than Europe or Japan, including two independent Spacelab missions, experiments on a free-flying platform (LDEF) deployed in 1984, the SPAR sounding rocket program (comparable to the German TEXUS program), experiments in drop tubes and drop towers, and participation in the German Spacelab-D1 mission.

Commercial users can obtain access to this NASA know-how through various means [104]:

- **NASA Databases.** NASA has a very extensive network of databases for all aspects of microgravity utilization [136]. Of these, the Aerospace Database, operated by AIAA, is the online version of two printed NASA publications: International Aerospace Abstracts (IAA) and Scientific and Technical Aerospace Reports (STAR). Our analysis of the major existing materials databases worldwide (Table 17) shows that the Aerospace Database is the largest source of information for a microgravity user interested in a comprehensive overview. This database is also accessible from Europe through the ESA-IRS* database service, but is not accessible from Japan. The mere existence

TABLE 17. *Usefulness of Databases as Information Sources for Microgravity Users*[a]

IRS Index[b]	Database	Microgravity	Transport phenomena	Surfaces, interfaces	Fluid dynamics	solidification	Structure of solids
1	AEROSPACE (1962 ff)	7,934	89,720	7,983	36,101	37,722	47,746
2	Chemabs (Vol. 66-106, 02)	1,110	119,906	23,523	10,615	166,089	88,450
3	METADEX (1969 ff)	191	33,671	6,175	3,016	59,563	65,642
4	COMPENDEX (1969 ff)	803	82,281	19,439	19,439	59,463	54,047
6	NTIS (1964 ff)	1,667	37,725	2,070	10,557	18,471	10,165
7	BIOSIS (1973, Vol. 32, 04; 83)	752	15,165	1,275	2,930	20,449	18,060
8	INSPEC (1971 ff)	1,024	84,749	14,868	27,726	89,140	49,522
9	Aluminum (1968 ff)	141	5,929	1,583	488	13,859	15,379
14	PASCAL (1973 ff)	480	114,345	5,924	20,057	198,041	102,218
48	Fluidex (1973 ff)	61	9,223	2,196	4,044	2,100	1,999

[a] Listed are the number of published articles on microgravity utilization in general and some other typical materials research fields that are important to microgravity experiments. Noteworthy is the Aerospace database with the largest number of articles on microgravity.
[b] IRS = Information Retrieval Service.

*IRS: Information Retrieval Service.

of the NASA database does not give U.S. users any advantage over European users. However, through the ten Industrial Application Centers (IAC) within its Technology Utilization Program (TUP), NASA provides the users with information retrieval services for computerized searches of the NASA database and more than 600 other databases [104,137].

- **Technical Exchange Agreement.** Under this agreement, NASA and a company agree to exchange technical information and cooperate in the conduct and analysis of ground-based research programs. The company funds its own participation, while gaining access to and results from NASA personnel and facilities. This results in intensive information exchange through cooperation, similar to the proposed joint projects in Germany.

- **Industrial Guest Investigators.** Researchers from user companies can work as guest investigators at NASA facilities at company expense and thus become familiar with the potential and limitations of microgravity. With this program, NASA is also building a base of technical advocates within the companies.

- **The Centers for Commercial Development of Space** (described in the next section) also play an important role in technology transfer. The CCDSs link industrial firms and universities with limited financial support from NASA. Technology is transferred via contacts with experienced experimenters.

The above comparison of these countries shows that the elements of technology transfer in Europe and in the United States are generally similar despite individual distinctions, except that technology in the United States is available from a single source (NASA), thus facilitating access. With Germany's depth and breadth of microgravity expertise, it would benefit from increasing efforts to make this expertise available more widely to future users. The Japanese approach may demonstrate that such systematic acquisition and distribution of information could plug large gaps in its own microgravity technology base.

INSTITUTIONS PROVIDING TECHNICAL AND ADMINISTRATIVE SUPPORT TO MICROGRAVITY USERS

During past space flight missions, the experimenters, the space industry, and official agencies have come to recognize the extensive activities involved that require special knowledge in the interactive areas between space flight providers (ESA and national space agencies), space hardware manufacturers, and microgravity users. These activities and the required expertise to perform them were outlined in Chapter 10. The partial lack of available services in Europe until the Spacelab-D1 mission is now covered by a large and growing number of private and public institutions [9,138]. These institutions are being established within the intersecting lines of activities by all participants in the space ventures.

Experimenters from European universities have established **user centers** to provide potential new users with technical support in conducting the microgravity experiments. These centers, which are mainly in Germany in keeping with the large German share of microgravity missions, generally receive seed money from the government, but are subsequently expected to be self-sustaining. Table 18 lists these centers.

INTOSPACE GmbH, based in Hannover, West Germany, was established to link microgravity users and the institutions providing space services.

INTOSPACE's first space project was the Cosima protein crystal growth experiment package launched on August 5, 1988, on a Chinese Long March-2 booster and attached to a reentry capsule, which was recovered on August 13. The company is evaluating two other missions: Suleika for processing superconducting materials, and Casimir for studying catalyst material processing under microgravity [93]. INTOSPACE is also coordinating the Japanese and European private sector experiments on the German Spacelab-D2 mission. INTOSPACE itself does not offer technical support, but assists the user with administrative details by its links and contacts in the space community (also see the discussion in "Project Accountability and Oversight" at the end of this chapter).

TABLE 18. *European Centers Providing Technical Support to Microgravity Users*

ACCESS
    Aachener Centrum für Erstarrung unter Schwerelosigkeit
    Location: RWTH, Aachen, Germany
CPC-μg Aachen
    Centrum für physikaliche Chemie unter Schwerelosigkeit Aachen
    Location: RWTH, Aachen, Germany
MATLAB-MUSC
    Materials Laboratory of the Microgravity User Support Center
    Location: DFVLR, Köln, Germany
BIOLAB-MUSC
    Biology Laboratory of the Microgravity User Support Center
    Location: DFVLR, Köln, Germany
WIB
    Weltrauminstitut Berlin
    Location: West Berlin, Germany
ZARM
    Zentrum für angewandte Raumfahrttechnik und Mikrogravitation
    Location: Bremen, Germany
BIOTECH-μg
    Zentrum für Biotechnologie unter Schwerelosigkeit
    Location: University Würzburg, Würzburg, Germany
MARS
    Microgravity Advanced Research and User Center
    Location: Technical University, Naples, Italy

Source: Reference [138].

In the meantime, there is a recognizable trend in the space industry to offer technical services, namely, integration of experiments and flight systems, in addition to more "classical" activities. These newer services include standard subsystems designed to accommodate payload elements like GAPS [139] or MAUS for mounting in GAS cans. These subsystems provide a user friendly interface to integrate the experiment and to reduce the costs of integration by standardization.

The **DFVLR** is a large national space research establishment in Germany [43] that offers user-oriented services and facilities for

- Converting laboratory experiments to flight-qualified experiments. The DFVLR Institute for Space Simulation in Cologne can procure functional duplicates of experimental flight hardware to conduct trial runs of the experiment. These capabilities have been extended as part of the user center MUSC and are being offered to ESA.

- Preparing operating procedures for astronauts and integrating experiment operations schedules into the flight schedules.

- Processing the data during and after the flight in the user control center within the German Space Operation Control Center (GSOCC) in Oberpfaffenhofen, where an extensive infrastructure is available for the user to control the operation and progress of the experiment.

These services tested well on the Spacelab-D1 mission. However, such services are not particularly oriented toward the interests of commercial users; for example, the control center (GSOCC) includes no mechanisms for securing experimental data during transmission to the user to protect proprietary interests.

At the DFVLR Institute for Space Simulation, ESA has also established a Microgravity User Support Center (MUSC).

In the United States, NASA provides technical support to microgravity users by the following mechanisms:

- The NASA field centers have a large inventory of development and test facilities for space-related hardware [74]. These facilities are available to the commercial user at nominal rates.

- NASA's Wallops Flight Facility in Virginia is available to entrepreneurial firms for flight testing and launches of commercial vehicles. The first agreement to this effect was signed in 1987 with Space Service, Inc., for launch of the Conestoga vehicles [104].

- At the Lewis Research Center in Cleveland, a Microgravity Materials Science Laboratory (MMSL) was built in 1985 [140]. The laboratory has available functional flight duplicates of experimental apparatus that can be used to develop flight-qualified experiments and conduct ground-based reference experiments. The private user can use the MMSL facilities under a TEA or through the IGI program. NASA subsidizes MMSL operations so that the prices charged for their use (mainly for proprietary work) are considerably below actual operating costs.

- NASA assumes integration of the experiment in the flight facility. Also, under certain conditions, NASA is prepared to modify its available equipment to meet the customer's requirements.

- NASA has established 16 Centers for the Commercial Development of Space (CCDS) [141]. A major goal of the CCDSs is to encourage commercial microgravity utilization. This is achieved by providing technical support and conducting space-based joint projects with industry; the 16 centers now have 119 industrial affiliates [104]. Each CCDS is a nonprofit consortium of universities, research institutes, industrial firms (users), and government, with either the host university or the host research institute managing the Center. In addition to their financial contribution (valued at $12 million in 1987 [142]), the industrial members also propose the research projects to be pursued and perform some of the preliminary scientific and technical work in their own laboratories (as in-kind contributions). The CCDS staff is responsible for the space-related part of the projects (experiment preparation, flight qualification, and integration into flight facilities). The Centers also assist in JEA negotiations with NASA. Flight opportunities aboard the Shuttle and sounding rockets for the Centers are handled on the same basis as other NASA JEAs, but involve some payment from the Centers. The CCDS programs are at least partly group projects in which several firms are involved in one specific project.

Table 19 lists the CCDSs involved in microgravity utilization. Each Center is funded by NASA at $750,000 to $1 million per year for five years, after which the Centers are expected to be self-sustaining. Since initiation of the program in 1985, CCDS accomplishments include over 600 drop tube and drop tower experiments, 21 aircraft flight experiments, and involvement in five Shuttle flights, including the first flight after resumption in September 1988 [104].

TABLE 19. *CCDSs in Microgravity Utilization*

Battelle Center for Advanced Materials
Columbus, Ohio
- Materials research under microgravity

University of Alabama, Huntsville, Consortium for Materials Development in Space
Huntsville, Alabama
- Materials research under microgravity

University of Alabama, Birmingham, Center for Macromolecular Crystallography
Birmingham, Alabama
- Materials research under microgravity

Vanderbilt University Center for Space Processing of Engineering Materials
Vanderbilt, Tennessee
- Materials research under microgravity

Clarkson University Center for Development of Commercial Crystal Growth in Space
Potsdam, New York
- Materials research under microgravity

University of Houston Center for Space Vacuum Epitaxy
Houston, Texas
- Materials research utilizing the vacuum of space

University of Wisconsin Center for Space Automation and Robotics
Madison, Wisconsin
- Development of robots for microgravity experiments
- Studies of automated mining on the moon

Case Western Reserve University Center on Materials for Space Structures
Cleveland, Ohio
- Structural materials under microgravity

In addition to NASA, several private companies and universities have identified facilities and services to support other users. These include payload development and processing (Astrotech, Wyle Laboratories, 3M, ITA, PSI), space hardware qualification and testing (Wyle Laboratories, Ford Aerospace), and turnkey operations to conduct microgravity experiments (PSI, ITA) [74].

In Japan, no institutions have as yet been formed for external technical support of commercial users, mainly because of limited Japanese experience with microgravity utilization. The Space Technology Corporation (STC), which is coordinating Japanese participation in the German Spacelab-D2 mission, describes one of its tasks as "performance of ground-based accompanying research to the flight program." The organizational chart of STC, however, shows only the laboratories of the member companies being used as the technical facilities for development of the space-based projects [99,100].

The Japan Space Utilization Promotion Center (JSUP) provides equipment for microgravity experiments.

STC and JSUP, as well as the Institute for Unmanned Space Experiments with Free Flyers (USEF), support their member companies in administrative matters, expressly including technology acquisition, and in negotiations with the mostly foreign suppliers of flight opportunities. However, comparison with European organizations, such as INTOSPACE, is not easy since these Japanese institutions represent a form of self-organization of user companies. Because of the often very large government financial contribution to these institutions (70 percent in the case of STC), however, this can also be considered a form of government administrative support.

By international comparison, potential microgravity users in Germany can find a diverse and extensive range of external support services. These services have resulted mainly from the individual initiatives of experienced space users and the space industry, and are the first fruits of the very strong involvement of Germany in microgravity. Even if federal institutions (DFVLR) are involved and private initiatives still have to be supported financially, the emerging scenario is of a large support base clearly oriented to the private sector.

The developments in the other ESA countries lag far behind those in Germany. Compared with the United States, Germany, in turn, lacks an overall conceptual framework (such as the U.S. CSP plan) within which the multitude of individual initiatives can be organized.

Even in the United States, the emphasis on private initiatives is very strong; for example, the CCDSs are reviewed annually for their success in obtaining new users as members, among other things. However, because all space authority is vested in NASA, NASA still retains a strong coordination function across the spectrum of commercial space activities. In Europe authority is divided between ESA and the national space agencies.

EXPERIMENTAL OPTIONS FOR RESEARCH IN MICROGRAVITY

Microgravity research can be conducted on orbital and suborbital flights, high-altitude balloon flights, parabolic flights in aircraft, in drop tubes and towers, or using ground-based simulation of individual effects of gravity. In the following sections, these experimental options will be described briefly, followed by analysis of the experimental options available in the individual countries. Detailed descriptions of each of these facilities are given elsewhere [74].

Basic Types of Experimental Options Under Microgravity

Spacelab. Spacelab is a reusable, manned laboratory attached to the Shuttle payload bay for conducting experiments under microgravity. Mission duration is 5 to 10 days. The presence of humans on the laboratory makes Spacelab experimental programs very flexible; experiments can be performed in which human intervention is expedient, perhaps because complete automation would be costly and technically risky.

On the other hand, human presence limits the quality of microgravity (g-level) and restricts the experiments to materials and processes with negligible hazard potential due to the strict safety requirements aboard the manned spacecraft.

The experimental facilities are predominantly multi-user facilities in which many different experiments are conducted in the same facility, thereby limiting the operational margin in the design of the experiments. Hence, Spacelab is particularly suited to R&D work under microgravity, whereas for actual production in space, automated, project-specific equipment aboard unmanned platforms would be preferred.

The Spacelab module is owned by NASA. It is designed to be launched into, maintained in, and returned from space by the Space Shuttle. Through the end of 1988, Spacelab had been employed four times of which one flight was under ESA direction (FSLP, 1983), two flights were under NASA control (Spacelab-2 and 3), and one was conducted for Germany (the D1 mission, 1985). The next mission is scheduled for June 2, 1989 [27].

Because Spacelab missions are large and expensive, a series of national missions can only be flown every few years. Besides the nationally controlled missions (the German D-series missions or the planned Japanese mission SL-J for FMPT), plans are being made to conduct international missions under the International Microgravity Laboratory (IML) program [143]. A wider sharing of the costs for Spacelab in the IML missions will permit more flights.

Free-Flying Platforms (Free Flyers). Typical examples of this experimental option are the American LDEF, the ESA program EURECA, and the planned Japanese free flyer (USEF). The most important characteristics of free flyers are described briefly using EURECA as an example.

EURECA will be launched on the U.S. Space Shuttle, will remain free flying in space outside the Shuttle for nearly nine months, and will then be retrieved on another Shuttle flight for return to Earth. The system is unmanned. As a rule, the experimental facilities are multipurpose and operate automatically. Thus, EURECA provides an opportunity to conduct long-duration experiments that require much higher microgravity quality.

Get Away Special (GAS). This refers to hollow cylindrical canisters, flown on the Shuttle, in which automatic self-operating experiments are mounted. The disadvantages—the need for automation, limited capacity, and poor microgravity levels—are far outweighed by definite advantages:

- More flights. Each Shuttle flight can carry GAS canisters, allowing a reflight frequency for GAS cans of nearly 9 to 12 per year.

- Lower flight costs. NASA charges between $3,000 (27 kg payload) and $10,000 (90 kg payload) for flying the GAS canisters.

In various countries, microgravity programs have been developed based on GAS options: either one GAS canister per experiment is used, such as MAUS in Germany, or to increase the overall capacity, several canisters are joined together, such as Interconnected GAS, GAS Bridge, MPESS, and Hitchhiker.

Space Station Freedom. The individual elements of S.S. Freedom are currently under review. As far as microgravity is concerned, three types of elements can be identified [66]:

- Pressurized module. This element, practically a Spacelab, is attached to the mechanical structure and the utilities of the S.S. Freedom. The difference here is only in the longer, essentially unlimited, duration of the mission.

- Co-orbiting platforms. This element is the EURECA platform.

- Columbus Free-Flying Laboratory. This element is also similar to Spacelab, but is not attached to S.S. Freedom. It combines the advantages of the other elements: various tasks can be performed manually by the astronauts; and in the unmanned phases, very high microgravity levels can be achieved for long experimental durations.

Sounding Rockets. Sounding rocket missions, such as the U.S. SPAR program, the German TEXUS program, or the MASER program of ESA (initial flight in February 1988), are suborbital flights. The duration of weightlessness is several minutes (6 minutes on TEXUS, 4 minutes 20 seconds on SPAR, 6 to 8 minutes on MASER), which means only short duration experiments can be performed. The mission frequency for TEXUS has been increased from once per year (two rockets launched within a few days) to twice per year (for a total of three launches per year).

Research Aircraft (Parabolic Flights), Balloon Flights, and Drop Towers and Tubes. These experimental options provide microgravity durations from several seconds to one minute. This can be adequate for short experiments or preliminary experiments to prepare for larger projects, and for testing individual experimental components or phases of an experiment. The advantages are very low cost, high reflight frequency, rapid access, and simple experimental assembly and a correspondingly smaller space-related preparatory effort. These options provide the commercial user with the means to make a cautious entry into and progressive involvement in microgravity utilization.

Simulation of Microgravity Effects. Except in free fall, gravity cannot be excluded on Earth and simulation of microgravity is basically not possible. But in certain cases, individual effects of gravity can be excluded by specific experimental methods. For example, solids or liquids can be levitated by electrical, magnetic, or acoustic forces. As in a microgravity project, the levitated materials can be studied without the disturbing effects from contact with container walls (containerless processing).

This type of experimental method is also useful to lower the inhibition threshold for commercial microgravity users: if containerless processing shows potential for commercial use, the process can be applied immediately to manufacturing and production on Earth. And if from the commercial usage a demand for improved boundary conditions arises, then projects can be initiated in a true microgravity environment.

Usefulness of Various Types of Flight Missions for Commercial Purposes. A specific preference of commercial users for one of the types of flight options is not evident; commercial utilization is still not prevalent enough to exhibit such a preference. As discussed above, commercial utilization will depend on the availability of a wide spectrum of diverse flight options for different kinds of projects.

GAS canisters, sounding rockets, parabolic flights, balloon flights, and drop tubes and towers offer a low-cost advantage for a single experiment, but the cost per kilogram of payload is very high, and even higher per kilogram-hour (see "Cost and Financing" in Chapter 11). These options are suitable for research work, but not for materials production, and are especially important in the initial feasibility phase of the project. Should microgravity prove to be useful in this phase of the project, then longer experimental times in higher quality microgravity (low g-levels) will be required for detailed studies, or even for production—a situation that will favor the use of free flyers like EURECA. Because Spacelab cannot offer the flight duration and microgravity levels achieved with the free flyers, it can be considered primarily for use in industrial research programs.

TABLE 20. *Types of Space Flight Programs and Providing Organizations for Various Programs with German Participation*

Experimental options and programs	Providing organization(s)	Manned or unmanned
Parabolic Flights (KC-135)	NASA (ESA, BMFT)	Manned
MIKROBA	DFVLR/BMFT	Unmanned
TEXUS	DFVLR/BMFT	Unmanned
MAUS	DFVLR/BMFT	Unmanned
SPAS	DFVLR/BMFT, ESA, NASA, MBB-ERNO	Unmanned
Spacelab	DFVLR/BMFT, ESA, NASA	Manned
EURECA	ESA	Unmanned
Columbus	ESA	

TABLE 21. *Duration and Quality of Microgravity (maximum) for Various Kinds of Experimental Microgravity Programs*

Experimental programs	Maximum microgravity duration	Average g-level
Parabolic flights	20s to 30s	10^{-2} g
MIKROBA	60s to 70s	10^{-3} g to 10^{-4} g
TEXUS	6 min	10^{-4} g
MAUS	up to days	10^{-4} g to 10^{-6} g
SPAS	10 days	3×10^{-4} g (with extended periods of 10^{-3} g)
Spacelab (shuttle)	10 days	10^{-4} g to 10^{-6} g
EURECA (free flyer)	6 months	10^{-4} g to 10^{-6} g
Columbus (space station)	permanent	to be determined

Structure of the Available and Planned Experimental Options for Research Under Microgravity

In Europe, Germany offers suborbital and orbital experimental options within national programs and participates in other options offered by ESA. Tables 20 and 21 give a brief overview of these options. Thus, users in Germany can choose among a very wide selection of experimental options.

Experience to date with these options and a better understanding of "typical" requirements of the experimenters have revealed gaps in this spectrum of experimental opportunities:

- The number of TEXUS flights is limited.

- Longer microgravity durations are needed than those available with sounding rockets like TEXUS and MASER.

- The power capacity of the single GAS canisters in the MAUS program is insufficient.

Modification of the known flight options is presently being considered to close these gaps: interconnected GAS and other methods to combine several GAS canisters, or to increase power for GAS canisters such as Hitchhiker-M, Hitchhiker-G, MPESS, from the Shuttle; larger TEXUS rockets with microgravity durations up to 15 minutes; a third TEXUS flight per year [144].

Another impetus for these changes has been the hiatus in Shuttle operations. The Challenger accident clearly revealed Germany's total dependence on the U.S. Shuttle for orbital missions such as Spacelab and EURECA. (Both missions have been delayed, Spacelab-D2 until December 1991 and EURECA until August 1991 [27].) Microgravity projects are currently feasible only with the suborbital programs (TEXUS). The following new programs have also been proposed:

- ORBIS is a German program to return orbital autonomous experiments under microgravity. In this program, the recoverable capsule Raumcourier or Revex for experimental microgravity payloads will be deployed. The capsules have the necessary facilities for re-entry into the Earth's atmosphere and subsequent landing, and can be launched by ELVs (including Ariane) into Earth orbit [145].

- TOPAS is a joint German-Italian venture in which microgravity payloads will be launched into Earth orbit on Scout rockets from the launch facility on San Marco Island off the coast of Kenya. The payloads will later be returned to Earth in a retrievable capsule such as Raumcourier and Revex. Flight durations are largely limited to a few days [145].

In this discussion of alternative recoverable systems designed to create a broad and secure base for microgravity ventures, two other plans will provide greater European autonomy:

- As a supplement to Columbus as the European contribution to Space Station Freedom, ESA is also making efforts to achieve independence from the U.S. Space Shuttle with its own manned system Hermes to be launched on Ariane-5.

- Germany has established bilateral cooperation with China and the Soviet Union for transport of payloads into space, a path long ago taken by France with experiments on the Soviet ELVs (Proton) and the MIR space station.

Experimental options are clearly diversifying. The process of further diversification and quantitative extension of the possibilities is ongoing. Even in the new programs being developed by ESA, leadership has again been assumed by Germany.

The other European countries, with the exception of the French bilateral ventures and the German-Italian TOPAS project, are participating in the ESA flight missions.

In the United States, NASA offers the following orbital space flight opportunities with experimental facilities similar to those available in Europe [74]:

- Spacelab and Spacehab (1991 operation)

- LDEF (similar to EURECA)

- GAS canisters, including MPESS and Hitchhiker

- Materials Science Laboratory (MSL)

- Space Station Freedom (planned).

Individual self-contained experimental apparatus, which is too large or too complex for integration into a GAS canister, can also be flown on the Shuttle. This provides an additional option for U.S. users of manned flight with high reflight frequency. An example of this kind of utilization is the installation in the Shuttle middeck of an electrophoresis unit for separation and purification of mixtures to produce pharmaceuticals.

Two additional options are also being evaluated. The ISF and CDSF are commercially developed man-tended free-flyers. Proposals for deployment of these facilities are being reviewed by the National Research Council, which is expected to complete its review in May 1989 [146]. In the suborbital and ground-based regions, NASA's experimental options include drop tubes and drop towers, parabolic flights with KC-135 and F-104 aircraft and the Learjet, and experimental opportunities on sounding rockets. NASA is also considering a new rocket that can fly payloads with microgravity durations of up to 20 minutes [144].

The American Spacelab missions carried fewer experiments per flight than the European FSLP and D1 flights. This allows for greater allocation of power, weight, size, data capacity, and astronaut time to each experiment, but also increases the share of the costs for each experiment.

Japan has its own sounding rocket, TT-500-A, to conduct an independent suborbital program. Until 1988, six flights had been conducted. Japan is at present participating exclusively on experimental options of other countries. The First Materials Processing Test (FMPT) was planned for a dedicated Spacelab flight (SL-J) in 1988, but this has been postponed to July 1991 because of the Shuttle delay [27].

A diversification of flight options and programs, as in the United States and above all in Germany, is not yet evident in Japan. However, Japan is planning the construction of a 500-meter drop tube in Hokkaido to be operational in 1990. Other plans include two drop tubes in Gifu Prefecture and Sagamihara City near Tokyo, and the use of prototype H-II for suborbital flights; the latter is about one-fifth the size of the final H-II rocket [100].

Overall, the orbital microgravity programs of all countries considered still very largely depend on the Space Shuttle. (This situation will change in the mid 1990s when free flyers are launched and recoverable capsules begin to be deployed.) This gives the United States political dominance over the ESA countries and Japan in microgravity utilization and in all the related implications for commercial utilization—control of access, proprietary rights, patent issues, and, not least, the potential to set prices for flights to deter competition.

The unavailability of the Shuttle from February 1986 until September 1988, however, has also shown that this dependence is technically very risky. The bottleneck in microgravity research (NASA's secondary payload backlog includes literally hundreds of experiments [146]) has occurred because of the lack of alternative launch systems and the lack of retrieval capacity. In launch capacity,

ESA with the Ariane vehicle, despite the recent hiatus in Ariane operations from May 1986 to September 1987, does have a capable and competitive (in other space segments) system, which could also be adapted to launch microgravity payloads. Efforts to use launch services in other countries (i.e., U.S.S.R. and China) are being made as mentioned; the only operational reusable system, however, is the Space Shuttle. Even the Soviet systems have problems with return of payloads as discussed earlier. Hermes, HOTOL, and SÄNGER are being conceived in Europe as reusable systems. Recently, the ESA Council gave its approval to begin the design and development phases of Hermes [109]. Thus, the commercial user in Europe would have available an independent European launch and retrieval system in the future. The additional support from space transportation of other countries will make the microgravity utilization scenario technically that much more reliable.

In suborbital programs, Germany holds a very favorable position with TEXUS; however, it still lags the United States in independent parabolic flight opportunities and number of drop tube and tower facilities. A 42-meter drop tower is available at ONERA in Paris, and a 47-meter drop tower is operational in Grenoble. A 100-meter drop tower is under construction at the University of Bremen [145], and a 1100-meter drop tube is being planned in Camphausen near Saarbrücken, West Germany [147]. Here, MIKROBA is expected to bridge the gap. The extensive involvement of the private sector in development of advanced flight options is expected to lead to an increase in these options based on user needs.

PROJECT ACCOUNTABILITY AND OVERSIGHT

In this area, the European countries are at a slight disadvantage, compared with the United States and Japan, because the transfer of some tasks and authority to a supranational organization (ESA) creates additional lines of intersection. Thus, it is particularly important for Europe to develop management models for microgravity projects, specifically with the needs of commercial users in mind. The model in Figure 11 [14] shows a typical scenario with the lines of interaction between the major participants in a microgravity project using ESA flight opportunities. The key participants are the experimenter (user), the space hardware manufacturer, ESA, and usually national financial or supporting institutions.

Such a complex scenario will not be attractive to potential microgravity users. The user can easily see the enormous time and financial effort that will be required to manage the project and to coordinate external contributions to the project.

In Europe, INTOSPACE was founded in 1985 under German leadership. Among its other goals is the "promotion, initiation, and support for industrial space activities in the area of weightlessness, such as research, development, and commercial product ventures" [93]. One main function of the company is coordination, planning, and consultation in scientific and technical areas, including seeking flight opportunities, financing space ventures, and legal and contractual matters. This is a major step in helping to relieve the user of coordination tasks. However, such support will have a significant impact on the project costs, since INTOSPACE provides its services on a commercial basis. At the same time, this impact will be felt only if

- Future interaction scenarios become considerably more focused and simplified, as compared with the scenario shown in Figure 11

- Sufficient experience has been gained using model scenarios for managing a commercial microgravity project

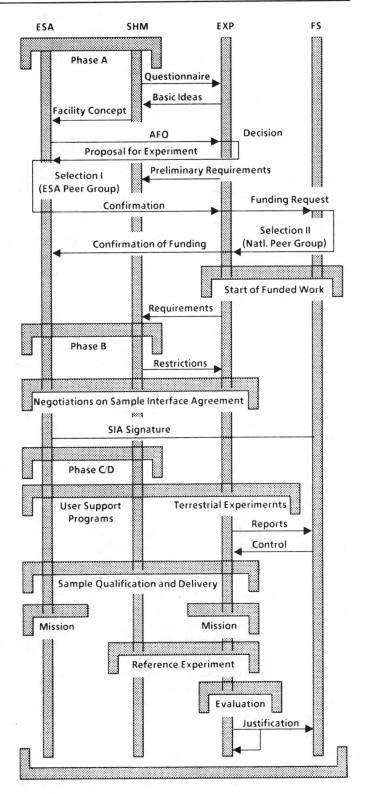

FIGURE 11.

Typical Scenario for Management of a Microgravity Experiment in Europe

- Uniform selection criteria exist for the experiments within ESA and within the national space agencies

- Support programs utilized by the user are modified to fit the overall management scenario for the project.

INTOSPACE is basically a European company; the shareholders are from each of the European countries considered here. The company offers its services worldwide. Given its founding partners, Aeritalia and MBB-ERNO, the distribution of shares (Germany at 45.7 percent, Italy at 26.5 percent), and its headquarters in Hannover, it is under strong German and less strong Italian influence. An independent French firm, Novespace, established on July 8, 1986, emphasizes spin-offs from space technology. Novespace also acts as an agent for cooperative projects between space companies and nonaerospace users. Novespace functions are in some respects identical to those of INTOSPACE [148].

In programs utilizing national flight opportunities, the various lines of interaction in Figure 11 are simplified. This gives countries with large national space flight programs (in Europe, particularly Germany) a slight advantage over countries participating only in international missions.

In the United States, the responsibilities and rights of the partners in a joint microgravity venture with NASA are defined in the JEA between the user and NASA. The user has to deal only with a single organization for project discussions and contract negotiations. Despite this relatively uncomplicated structure, coordination problems still occur. Until recently, JEA negotiations with NASA often took up to two years until an agreement could be signed. Such lags can mean the difference between continued operation and failure for a private company. NASA has developed new processes for proposal review, evaluation criteria, and review process guidelines that should shorten decision time on even moderately complex JEA proposals to four to six weeks [149].

The collaboration between commercial microgravity users and NASA does not begin only with negotiations for a JEA. NASA offers a much broader range of services with initiatives to support commercial users. These initiatives have been discussed in detail earlier and compared with similar initiatives in other countries. NASA has developed a comprehensive framework through the Commercial Use of Space Policy (CSP plan) within which all these initiatives can be logically organized. In contrast to all other countries, the American space user can obtain a complete overview, not only of political principles, but also of individual initiatives from the CSP plan, on which users can organize their own project plans. The important initiatives in the CSP plan are summarized in Table 22. A more detailed description of the political principles and the initiatives is given elsewhere [131].

As noted earlier, a number of private U.S. companies provide facilities or turnkey services to support commercial users. Here again, the user has to deal only with a single organization providing these services.

Since Japan has conducted very little independent microgravity research, it has very little experience in managing such ventures. But there are two reasons to expect that project management models being developed as part of the Japanese preparations for their experiments will be convenient and manageable for the commercial user:

- In Japan the interface between the project partners has traditionally been very important. For space projects, there are existing models and experiences available from other fields such as semiconductor technology.

- In Japan, industry has been extensively involved in microgravity utilization from the beginning, in complete contrast to Europe and the United States. This will certainly ensure the development of working models that are favorable to private industry from the start.

TABLE 22. *Support Initiatives for Commercial Utilization of Microgravity in the United States*

Technical Support:
Access to NASA's ground-based experimental facilities as part of
- TEA (Technical Exchange Agreement)
- IGI (Industrial Guest Investigator Agreement)
Disseminate widely information on space technology through the TUP
Payload integration assistance
Perform independent experiments in basic research areas

Financial Support:
Provide low-cost space flight opportunities through
- JEA (Joint Endeavor Agreement)
- GAS (Get Away Special) Program
Partial subsidy for costs of payload integration
Direct financial support of projects through SBIR program
Limited purchase guarantees for space products

Institutional Support:
CCDS (Center for Commercial Development of Space)

Legal Initiatives:
Protection of property rights, patent rights, data, and company proprietary information under
  JEAs

CHAPTER

14 Comparison of Existing Conditions in Specific Countries

In this chapter, the discussion will focus on an international comparison of the degree to which the conditions required for commercial microgravity utilization (delineated at the end of Chapter 10) are met by the framework conditions described in Chapters 11, 12, and 13.

MARKET CONDITIONS

The market situation in the material sciences areas (production of pharmaceuticals also belongs here) shows no qualitative differences. U.S. businesses may indeed operate within a larger domestic market. On the other hand, high-technology, value-added products (and only such products will be manufactured in space) are usually sold worldwide, which means companies in all countries operate within the same market. Differences in the market potential, therefore, result more from trade policies and strategies (protectionism, tariffs, duties) in the individual countries. These issues are not unique to microgravity utilization for materials processing.

In the biological and life science disciplines, a distinction must be made between microgravity project results for use in space and those for use on Earth. Results for use on Earth face the same situation as in the materials sciences areas. Results for use in space are part of a feedback effect: increasing use of space creates a need for increased biomedical research under microgravity. Ambitious U.S. programs like manned missions to Mars could eventually create a slight advantage for U.S. companies, but such an advantage would be relatively small.

COST FACTORS

The costs of utilization of microgravity are determined by three main factors:

- Development, construction, and operation of the flight system, including launch

- Tailoring the flight program to the objectives of the respective microgravity project

- Quality of the supporting infrastructure.

National differences are most likely to be found in the last two factors. These will be discussed in more detail.

FINANCIAL SUPPORT

All of the countries considered are still willing to assume the bulk of specific space-related additional costs to the user in a microgravity project. Only NASA has begun (recently) to demand financial repayment of its previous support from any commercial profits of a microgravity venture as part of the Space Systems Development Agreement (SSDA). Thus, only the United States has created a financial instrument in which the government no longer covers the additional costs, but only reduces the financial risk of space-based ventures. This mechanism will be effective at the stage after the first private sector microgravity projects have been commercially successful. The JEAs are model agreements for the type of financial support mechanism on which future users can orient themselves. By contrast, such clearly defined financial models are lacking in Germany and elsewhere.

In accordance with the varying space promotion policies of individual countries, there are differences in the financial support available for ground-based laboratory research in support of a space project. These differences are, however, not specific to microgravity and do not, therefore, influence a company's decision on whether a commercial venture should include microgravity or be limited to terrestrial options. This neutrality of different promotion policies is based on the assumption that access to available resources of support for the microgravity venture should neither discriminate against nor favor the venture. Favorable conditions would eventually encourage more industrial users to perform microgravity experiments, but these would again be mainly academic projects, not commercial ventures. All financial support mechanisms, unless specially designed to support microgravity ventures, presently discriminate against microgravity utilization because they are irreconcilable with the special boundary conditions of microgravity projects. The necessity to tailor existing programs to these conditions is vital to Germany, which, like Japan and the United States, does have a broad spectrum of financial support programs.

GUARANTEED ACCESS

To guarantee access to space, four aspects must be considered:

- Long-term assurance related to cost evolution
- Political guarantees of flight opportunities
- Technical reliability of flight opportunities
- Organization and regulation of access and reflight frequency.

On the first point, Germany is presently at a disadvantage, compared with the United States and with France, in orbital flight opportunities since Germany does not have an autonomous launch and retrieval system for microgravity payloads and it has developed few alternatives to the Space Shuttle, such as Chinese and Soviet flight opportunities (although additional alternatives are being developed, as discussed in the previous chapter under "Structure of the. . .Experimental Options").

France, too, has developed limited alternatives to fly microgravity payloads through bilateral cooperation with the Soviet Union and China. For suborbital flight opportunities, the situation is different since Germany has a reliable and capable system in TEXUS.

In the longer term, Germany is planning to secure a controlling position for access to orbital flight opportunities by its large contributions to the Columbus project and future European manned space transportation systems such as Hermes (and possibly HOTOL and SÄNGER). At the same time, this will also create competition in launch systems with increased launch capacity, which will give Germany viable alternatives to counteract any eventual unfair price competition from the United States. (This is highly unlikely in view of the increased competition in available launch capacity in the 1990s.) The resulting advantages—reasonable cost evolution and guaranteed access to space—can be considered a positive framework condition for the private sector. Thus, Germany is internationally also very well positioned with regard to access to orbital and suborbital flight opportunities.

Japan is in a similar situation as Germany and the other European countries. Sounding rockets for suborbital flights do exist, but are seldom used for microgravity payloads. The advanced Japanese launch vehicles (e.g., H-11 rocket) could be used in the future for microgravity payloads, giving Japan greater independence from the United States.

During the Shuttle hiatus, potential microgravity users realized that guaranteed access to space is the most critical issue for space-based microgravity utilization. The unilateral dependence on the Shuttle has been a major disadvantage for all countries involved and efforts have begun to develop technical alternatives for access to space.

In terms of organization, and in particular, the capacity and reflight frequency of individual missions such as the Spacelab-D series, Germany does not have a strong base to support significant commercial activity in microgravity utilization. This situation is a direct consequence of the complete dependence on the Shuttle and will change perceptibly only when the technical alternatives for access to space are fully developed. This is also true of the other European countries and, to a lesser extent, Japan.

Only the United States has a manned reusable space transportation system with a high reflight frequency, since each Shuttle flight, in addition to the few dedicated microgravity missions, can also be used for microgravity work. American industry has a slight advantage here. However, with the planned expansion of the TEXUS program and participation in the Spacelab-IML missions, Germany is keeping its options open to develop additional experimental capacity if demand increases even before Space Station Freedom begins operation.

Overall, Germany is still well behind the United States, although less so than France, in access to space. The planned new programs, however, will enable Germany to provide its commercial users politically and technically secured access to sufficient capacity for microgravity research in space.

LEGAL ISSUES

Legal issues are regulated in the United States by the Joint Endeavor Agreements. The first experience with protection of proprietary information for commercial users, such as the Shuttle experiments performed by 3M, is already available. By contrast, Europe and Japan lack well-defined models, which certainly results from the time lag in private sector involvement in microgravity activities in these countries. The gap will begin to close with the increased participation of private companies in the German Spacelab-D2 mission.

To place the European user industry on an equal status with U.S. industry will very largely depend on how the legal issues are resolved and regulated for S.S. Freedom. Negotiations are ongoing, so the situation cannot yet be evaluated.

TECHNOLOGY BASE

In international comparison, Germany possesses a very extensive base of knowledge in microgravity, as does the United States, with France, Italy, and Great Britain much farther behind. Japan has practically no independent expertise.

Knowledge of microgravity is broadly distributed in Germany, and the principal sources are young scientists at the universities. This is an excellent starting position for the gradual dissemination and transfer of this knowledge to private industry. In the short term, however, a significant part of this knowledge is being lost in Germany:

- Through a lack of effort to archive and organize existing knowledge in databases

- Because evaluation of the previous microgravity experiments is limited to scientific analysis

- Through the forced interruption in orbital flights during the Shuttle hiatus, combined with strong fluctuations in the pool of technical experts.

Exactly the opposite situation is developing in Japan: the lack of independent knowledge is compensated by systematic analysis and evaluation of previuous European and American microgravity projects. This "borrowed" knowledge is systematically disseminated to private industry. This transfer is easier in Japan since private industry is more involved in the early precommercial phases of microgravity utilization than in Europe or the United States.

In the United States, too, dissemination of microgravity knowledge and experience is receiving considerable attention. As in Germany, transfer of this knowledge to private industry is accomplished through cooperation between industry and experienced NASA personnel or university and other research scientists. Such cooperation is better organized in the United States through the CCDSs than Europe since industry membership is a binding prerequisite for financial support of the CCDSs by NASA.

In Germany the very first joint space utilization projects between industry and universities are being developed within the BMFT framework of initiatives for support of research.

TECHNICAL AND ADMINISTRATIVE SUPPORT

A multitude of institutions have been formed in Europe, Japan, and the United States to support commercial users. In Germany, centers for technical support (Centers of Excellence) are largely the result of private initiatives. Government control of these centers is less prominent than in the United States, although they do receive federal support, and many are under no government control pressure to obtain industry membership. Besides Germany, there is one such center in Italy.

Technical support centers do not exist as yet in Japan, but the user companies have organized several consortia. These consortia provide support predominantly in administrative matters and serve as conduits for government financial support. From them, however technical support centers can be quickly developed.

The United States provides technical support primarily through the CCDSs and the NASA field centers and offices. These organizations are subject to NASA control and provide such support largely within the framework of the CSP plan.

From this comparison, it is apparent that the user companies in Japan are self-organized with government support, whereas technical centers in Germany formed intially without industrial participation. On the other hand, the CCDSs in the United States are a synthesis of both these forms. NASA initiates and supports the CCDSs as technical centers, but links them to consortia from user industries.

These institutions provide user companies in the United States as well as in Germany with the necessary resources for technical support of an experimental project. However, differences in the functional details should not be overemphasized since the development of this infrastructure is not yet complete. The fact that, in Germany, such centers are formed largely from strong private sector initiatives guarantees the necessary dynamism for their further development. However, the German centers and the Japanese consortia have had less success than the CCDSs in bridging the gap between user needs and the potential of microgravity. Because the German centers do not attract much interest from potential private users, there is the risk that the technical support and services offered by these centers will once again be oriented primarily toward university researchers and insufficiently toward industrial users. The Japanese consortia do lack a range of technical services to offer, but the starting position in Japan is certainly more favorable than in Germany for further development of industry-oriented technical infrastructure: the lack of technical infrastructure can be (and currently is) compensated for by acquisition ("purchase") of technology in Europe and the United States; independent facilities patterned after selected successful facilities in Europe and the United States can be developed. By contrast, the reservations of European industry in the initial development phase of microgravity utilization will make it more difficult to further develop an industry-favorable infrastructure in Germany and in Europe as a whole.

This wait-and-see attitude of German and other European user industries emphasizes the need to acquaint industry with the extensive range of technical and administrative support services available throughout Europe. Here creation of a conceptual framework, like the CSP plan in the United States, within which the multitude of private support initiatives particularly in Germany can be consolidated, will be very helpful.

RESEARCH OPTIONS

Experimental options for research under microgravity are presently offered mainly by the United States, ESA, and Germany. As in the United States, Germany has a very diversified spectrum of flight opportunities available to commercial users. The only currently operational suborbital flight program among these is the TEXUS program. The process of diversification is not yet completed. In particular, the following programs with a major German share have been proposed and are being developed within Germany and within ESA:

- European contributions to the planned Space Station Freedom (Columbus)

- Extension of the sounding rocket program to longer flight durations

- New retrieval systems for microgravity payloads using recoverable capsules

- Modification of GAS canisters.

Germany and the ESA lag behind the United States only in the "small" experimental options, drop tubes and towers, and parabolic flights in aircraft. A limited number of drop tubes and drop towers exist in Europe, but operating experience with these facilities is very limited. Parabolic aircraft fights are available only in cooperation with NASA. Thus, there is lack of rapid and frequent access to inexpensive experimental options by which potential commercial users can initially evaluate the benefits of microgravity to a project.

The available space flight programs in Germany have been strongly oriented toward academic users. Reflight frequencies are too low, and the question of protection of proprietary information is still unresolved. The decisive breakthrough in reflight frequencies will come only with the deployment of S.S. Freedom. The issue of proprietariness, and its effects on such technical issues as qualification programs, safety philosophy, or data processing in the mission control center, could be addressed and solutions tested on future missions such as Spacelab-D2 and IML, which have significant private sector involvement.

The other European countries have access to ESA flight opportunities and, in general, to German flight programs. The space available for these countries is smaller, based on their smaller financial contributions. For the German industry, the national flight programs have the added advantage of guaranteed access, pricing policy, and legal security.

Japan is still dependent on external suppliers of flight opportunities, but is planning to develop autonomous flight programs.

PROJECT MANAGEMENT

In project management scenarios for industrial users, the United States has a clear lead. The CSP plan provides a framework for the interactions and interfacing of all project participants. The JEAs define the details. Here, the user has to deal only with a single partner (NASA or commercial providers): all services from flight opportunities to technical support to financing mechanisms and financial support come from a single source.

By contrast, the juxtaposition of ESA and the national space agencies already has resulted in several areas of friction, such as in microgravity programs in Germany and in remote sensing in France. The risk of increased friction is particularly high in Germany, where national flight programs compete directly with ESA programs.

A very complex interaction model, such as that shown in Figure 11, is unattractive for commercial users. Here a company such as INTOSPACE can assume all coordination responsibilities and relieve the user of project management activities for the various tasks and participants in the project. Even so, the suppliers of services must agree and coordinate activities among themselves to meet the requirements for commercial utilization.

The cost accountability of microgravity ventures is closely associated with the functionality of the project management model. The CSP plan and the JEA regulations help U.S. users with project management and clarify the respective roles of the partners. A private company providing these services assumes a role similar to that of NASA.

The multitude of individual initiatives in Europe for services, as discussed above, is essentially a positive development. However, without a conceptual framework to organize these initiatives, a very unclear and diffuse picture emerges. The functions and roles of the individual elements of the infrastructure (INTOSPACE, user centers, industry, national institutions, ESA) are not clearly defined. The services offered by the user centers vary widely (determined by their historical development), and users cannot easily assess their capabilities to support a commercial venture. Hence, users cannot be sure what specific assistance is available from which partner.

CHAPTER

15

Microgravity: Conclusions

The analysis of the existing framework conditions and comparison with the requirements for potential commercial utilization of microgravity has shown that the significant barriers to commercialization are not specific to a country, but are present in similar form in all countries. The general barriers include

- No market for microgravity products or results

- The absence of proven products

- Limited ideas for commercial microgravity projects

- The high cost of access to space

- The long duration of microgravity projects

- Long payback periods and high financial risk.

Some national differences exist in the initiatives developed to overcome these barriers, mainly in relation to

- The importance attached to microgravity *per se* or in relation to other space activities

- The scope and size of scientifically oriented (academic) microgravity programs

- The development of essential infrastructures

- The present involvement of industry in microgravity utilization.

Germany has always emphasized microgravity utilization, as expressed in the large variety of national programs and in the leadership role Germany plays within ESA programs. As a result of this strong involvement, microgravity research has a leading position in Germany. At the moment, Germany also has a very extensive base of technical and scientific knowledge. Even by comparison with the United States, Germany's position can be considered very good.

However, German knowledge is not well organized or efficiently disseminated, nor is that of other European countries, in contrast with the United States. Japan appears able to compensate for the lack of its own knowledge base by systematic analysis and transfer of current knowledge from Europe and the United States.

A further positive result of the extensive involvement of Germany in microgravity is the development of the infrastructure, which is also more advanced than in the neighboring European countries and Japan. Compared with the United States, however, Germany's conceptual structure of this infrastructure is less developed. Most affected are supporting institutions such as user centers and the services offered by federal agencies, whereas available flight opportunities are relatively plentiful. The essential elements of such a user support infrastructure are present in Germany. There is also no lack of private sector initiatives, as well as federal initiatives for further development of the existing infrastructure. But a conceptual framework for commercialization of microgravity, within which the existing elements and any new initiatives can be organized, does not exist. This makes the available infrastructure much too confusing to the commercial user.

The current limited involvement of industry in microgravity utilization and the simultaneous absence of direction that would be provided by such a conceptual framework makes it difficult for the emerging user centers in Germany and other European countries to orient their services toward industry. Here Japan is in a very favorable situation, since Japanese industry is also intensively involved in microgravity, even in its precommercialization phase.

The spectrum of flight opportunities in Europe, particularly in Germany, is certainly comparable to the American spectrum; in fact, the European spectrum is even more diversified. In the time before Space Station Freedom becomes operational, it is important for all countries to continue current and planned microgravity programs. In this regard, Germany's planned development of further flight options puts it in a very favorable position.

The willingness to achieve limited autonomy with new flight options, from the near total dependence on the Shuttle, enhances any country's commitment to space. This, in turn, increases the confidence of commercial users in the technical and political dependability of access to space, making the private sector more willing to become involved. By realizing these alternative options, and if the Columbus project and a manned space transportation system such as Hermes lead to development of a largely independent European utilization scenario with a significant German share, private companies in Germany will no longer be at a disadvantage, compared with companies in other countries, with respect to the framework condition "access to space."

Even if the precommercial involvement of industry is perceived as a framework condition to be provided by the government for microgravity commercialization, this involvement does affect the emerging infrastructure and the generation of ideas for future commercial ventures. Therefore, an important task in the current development phase is organizing the recognizable and developing interests of commercial users. Giving the topic of microgravity high visibility and value in the industry and organizing interested companies according to microgravity disciplines is an important step. This process of self-organization by discipline of potential user companies is developing in Japan, and is being encouraged in the United States by linking the microgravity user centers (CCDSs) to a consortium of companies. Government financial initiatives for microgravity research should be integrated into general financial support programs, as far as possible, and should require increasingly independent company financing as the project nears commercialization. Here the United States, with its policy initiative to provide special seed money for groups of companies with similar interests, appears to have created a promising mechanism for success (SBIR, CCDS). The United States clearly leads the European countries in this respect; Japanese industry is also very involved in space.

Besides seed money, governments have the option to promote precommercial involvement of industry by intensifying the dialogue with private industry and promoting industry-oriented microgravity demonstration projects. Too much pressure for commercial utilization, perhaps by advertising campaigns or even the use of too many financial incentives, will not achieve self-sustaining industrial involvement. Companies will be skeptical and critically analyze the purported benefits. Instead, the objective of this dialogue must be to increase the awareness of industry of the technical potential of microgravity and of the specific boundary conditions for such projects (flight facilities, project organization, technical and administrative infrastructure, costs). Corresponding efforts in the United States have profited from the better perception of the management scenario and from coordination of all activities through a single organization, either NASA or commercial service providers.

Compared with basic research-oriented microgravity projects, industrial-oriented demonstration projects must emphasize application of the results and commercially sound project management, including the experimental approach, resolution of legal issues, and cost and schedule control.

Based on this overview of framework conditions, Germany leads Europe in microgravity and has equalled the United States in technical maturity, even surpassing it in some areas. Inadequate organization, however, limits Germany's ability to apply its knowledge, thereby limiting this positive framework condition.

The situation in Japan is highly organized. Idea generation is ongoing in industry. An independent base of fundamental knowledge and experimental technology does not exist, but is being systematically acquired through cooperation with the United States and Europe, particularly Germany. Japan may close the gap very quickly.

The United States most effectively encourages the development of commercial microgravity ventures. Aside from the benefits of the reusable Space Shuttle, the potential for microgravity utilization in the United States is not much greater than in Germany. However, infrastructure coordination is greater, and efforts in the direction of commercial utilization of microgravity are more focused.

List of Acronyms

ACE. Army Corps of Engineers (United States)
ACR. active cavity radiometer (United States)
ADEOS. Advanced Earth Observation Satellite (Japan)
AID. Agency for International Development (United States)
AIAA. American Institute of Astronautics and Aeronautics (United States)
AIST. Agency for Industrial Science and Technology (Japan)
ANVAR. Association Nationale de Valorisation de la Recherche (France)
Ariane. European expendable launch vehicle (ESA; France)
ASEAN. Association of South East Asian Nations
ASI. Agenzia Spaziale Italiana (Italy)
AVHRR. advanced very high resolution radiometer (United States)
AVIRIS. airborne visible/infrared imaging spectrometer (United States)

BARSC. British Association of Remote Sensing Companies
BGR. Bundesanstalt für Geowissenschaften und Rohstoffe (West Germany)
BDPA. Bureau pour le Développement de la Production Agricole (France)
BMFT. Bundesministerium für Forschung und Technologie (West Germany)
BNSC. British National Space Centre
BRGM. Bureau des Recherches Géologiques et Minîeres (France)
BTMC. Bell Telephone Manufacturing Company (United States)

Carina. reentry vehicle for microgravity payloads (Italy)
CASS. Committee on Aeronautical and Space Sciences (United States Senate)
CCDS. Center for the Commercial Development of Space (United States)
CCIR. Comité Consultatif International des Radiocommunications
CCT. computer compatible tape
CDAS. command and data acquisition station
CDSF. Commercially Developed Space Facility (United States)
CEC. Commission of European Communities
CEDIS. Combined Environmental Data Information System
CENG. Centre d'Études Nucléaires de Grenoble (France)
CGMS. Coordinating Meeting for Geostationary Meteorological Satellite
CIPE. Comitato Inteministeriale Perla Economia (Italy)
CLS. Collecte Localisation Satellites (France)
CNES. Centre National d'Études Spatiales (France)
CNR. Consiglio Nazionale delle Ricerche (Italy)
CNRS. Centre National de la Recherche Scientifique (France)
COMSAT. Communication Satellite Corporation
COPUOS. Committee on Peaceful Uses of Outer Space (United Nations)
CSG. Centre Spatial Guyanais (France)
CSP. Commercial Use Space Policy (United States)
　　　 Center for Space Policy (United States)
CSPI. Commercial Space Policy and Implementation (United States)
CUPP. Columbus Utilization Preparation Program (ESA)
CZCS. coastal zone color scanner (United States)

DCP. data collection platform

DCS. data collection system

DFD. Deutsches Fernerkundungs-Datenzentrum (West Germany)

DFVLR. Deutsche Forschungs- und Versuchsanstalt für Luft- und Raumfahrt (now known as Deutsche Forschungsanstalt für Luft- and Raum fahrt) (West Germany)

DLR. Deutsche Forschungsanstalt für Luft- und Raumfahrt (West Germany)

DMSP. Defense Meteorological Satellite Program (United States)

DOC. Department of Commerce (United States)

DOD. Department of Defense (United States)

DOI. Department of Interior (United States)

DOT. Department of Transportation (United States)

DPC. Data Processing Centre

DRS. Data Relay Satellite (ESA)

DSN. deep space network

DWD. Deutscher Wetterdienst (West Germany)

ELA. Ariane launch complex (ESA)

ELDO. European Launcher Development Organization

ELV. expendable launch vehicle

EMSI. European Manned Space Infrastructure (ESA)

EOC. Earth Observation Center (Japan)

EOS. Earth Observation System (United States) Electrophoresis operation in space

EOSAT. Earth Observing Satellite Company (United States)

EPIC. Establissement Public Industriel et Commercial (France)

ERS. European Remote Sensing Satellite (ESA) Earth Resources Satellite (Japan)

ERSDAC. Earth Resources Satellite Data Analysis Center (Japan)

ESA. European Space Agency

ESMR. electrically scanning microwave radiometer

ESRIN. European Space Research Institute (ESA)

ESRO. European Space Research Organization

ETCA. Etudes Techniques et Constructions Aérospatiales (France)

ETM. enhanced thematic mapper

EURECA. European retrievable carrier (ESA)

EWIV. European Economic Interest Group

FAA. Federal Aviation Administration (United States)

FAO. Food and Agricultural Organization (United Nations)

FESTIP. Future European Space Transportation Investigation Program

FMPT. first materials processing test (Japanese Spacelab mission)

FSLP. first Spacelab payload (1983 ESA Spacelab mission)

GAF. Gesellschaft für angewandte Fernerkundung (West Germany)

GARP. Global Atmospheric Research Program

GAS. Get Away Special (United States)

GDTA. Groupement d'Intérêt Economique pour le Développement de la Télédétection Aérospatiale (France)

GIE. Groupement d'Intérêt Economique (France)

GIS. geographic information system

GMS. geostationary meteorological satellite

GNP. gross national product

GSOCC. German Space Operation Control Center

GTO. geosynchronous transfer orbit

Hermes. European manned space vehicle
HIRS. high-resolution infrared radiation sounder
HOPE. H-II Orbit Plane (Japan)
HOTOL. horizontal take-off and landing (British manned space vehicle)
HR-FAX. high-resolution facsimile

IAA. International Aerospace Abstracts
IAC. Industrial Application Center (United States)
IFP. Institut Français du Pétrole (France)
IGI. industrial guest investigator (NASA)
IGN. Institut Géographique National (France)
IML. International Microgravity Laboratory
IRDI. Institut Régional de Développement Industriel Midi-Pyrénées (France)
IRI. Instituto per la Riconstruzione Industriale (Italy)
IRS. Information Retrieval Service (ESA)
Indian Remote Sensing Satellite (India)
ISAS. Institute of Space and Astronautical Science (Japan)
ISF. Industrial Space Facility (United States)
ITA. Instrumentation Technology Associates (United States)
ITD. Institute for Technology Development (United States)
ITU. International Telecommunication Union

JEA. joint endeavor agreement (United States)
JEM. Japanese experimental module
JPL. Jet Propulsion Laboratory (United States)
JSUP. Japan Space Utilization Promotion Center

KC-135. NASA aircraft for parabolic flights (United States)
Key-TEC. Key Technology Center (Japan)

LDEF. NASA long-duration exposure facility (United States)
LEO. low Earth orbit
LES. Laboratoire d'Étude de la Solidification (France)
LFC. large-format camera
LFMR. low-frequency microwave radiometer
LIDAR. light detection and ranging
LR-FAX. low-resolution facsimile

MARS. Microgravity Advanced Research and User Support Center (Italy)
MASER. materials science experiment rocket (Sweden)
MAUS. Materialswissenschaftliche autonome Untersuchungen unter Schwerelosigkeit (West Germany)
MDAC. McDonnell Douglas Astronautics Company (United States)
MESSR. multispectral electronic self-scanning radiometer (Japan)
MIKROBA. Mikrogravitation mit Ballonen (West German)
MIPAS. Michelson interferometer for passive atmosphere sounding
MITI. Ministry of International Trade and Industry (Japan)
MLA. multilinear array (United States)
MLSA. modified launch services agreement (United States)
MMSL. NASA Microgravity Materials Science Laboratory (United States)
MOMS. Modularer Optoelektronischer Multispektraler Scanner (West Germany)
MOS. marine observation satellite (Japan)
MOU. memorandum of understanding
MPS. materials processing in space
MPESS. multipurpose experiment support structure

MRI. Mitsubishi Research Institute (Japan)
MSL. Materials Science Laboratory
MSR. microwave scanning radiometer (Japan)
MSS. multispectral scanner (Landsat, United States)
MTFF. man-tended free flyer
MUSC. Microgravity User Support Center (ESA; DFVLR, West Germany)

NASA. National Aeronautics and Space Administration (United States)
NASDA. National Space Development Agency (Japan)
NESDIS. National Environmental Satellite Data and Information Service (United States)
NOAA. National Oceanic and Atmospheric Administration (United States)
NPOC. national point of contact
N-ROSS. Navy Remote Ocean Sensing Satellite (United States)
NRSC. National Remote Sensing Centre (Great Britain)

OECD. Organization for Economic Cooperation and Development
OMB. Office of Management and Budget (United States)
ONERA. Office National d'Études et de Recherches Aérospatiales (France)
ORBIS. Orbital Flugprogrammen mit Rückführgelegenheit zum Betrieb autonomer Experimente in der Schwerelosigkeit (West Germany)
OSIRIS. oxide-dispersed single crystals improved by resolidification in space (West Germany)

PAPA. Ariane operations support program (ESA; France)
PEPS. Programme d'Evaluation Preliminaire des données Spot (France)
PRARE. precise range and range rate equipment
PSN. Piano Spaziale Nazionale (Italy)
PSI. Payload Systems Incorporated (United States)
PTT. Post, Telephone and Telegraph Administration

Raumcourier. retrievable space capsule (West Germany)
RESTEC. Remote Sensing Technology Center (Japan)
Revex. retrievable space capsule (West Germany)
RWTH. Rheinisch-Westfällische Technische Hochschule (West Germany)

SAC. Space Activities Commission (Japan)
SÄNGER. two-stage space vehicle (West Germany)
SAR. synthetic aperture radar
SBIR. Small Business Innovation Research (United States)
Sea-WiFS. sea-wide field sensor (United States)
SEP. Société Européenne de Propulsion (France)
SIR. shuttle imaging radar
SMMR. scanning multichannel microwave radiometer (United States)
SOASS. solar occultation absorption spectroscopy sensor (United States)
SPAR. Space Processing Application Rocket (NASA)
SPAS. shuttle pallet satellite (West Germany)
SPOT. Système Probatoire d'Observation de la Terre (France)
SRB. solid rocket boosters
SSC. Swedish Space Corporation
SSDA. Space Systems Development Agreement (United States)
SSM/I. sensor system microwave/imager (United States)
SSP. Space Station platform
STA. Science and Technology Agency (Japan)
STAR. satellite tracking and reporting
Scientific and Technical Aerospace Reports (United States)

STC. Space Technology Corporation (Japan)
STET. Società Finanziana Telefonica p.a. (Italy)
STIID. Scientific, Technological and Industrial Indicators Division (OECD)
STS. Space Transportation System

TASC. The Analytic Sciences Corporation (United States)
TDM. thousand German marks
TEA. technical exchange agreement (NASA, United States)
TEXUS. Technologisches Experiment unter Schwerelosigkeit (West Germany)
TIR. thermal infrared
TM. thematic mapper (Landsat, United States)
TOPAS. transport operation of microgravity payloads assembled on Scout
        (West Germany)
TOPEX. ocean topography experiment (France; United States)
TUP. NASA Technology Utilization Program (United States)

UNIDO. United Nations International Development Organization
USDA. U.S. Department of Agriculture
USEF. unmanned space experiments with free flyers (Japan)
USML. U.S. Microgravity Laboratory

VISSR. visible and infrared spin scan radiometer
VNIR. visible and near-infrared radiometer
VTIR. visible and thermal infrared radiometer (Japan)

WARC. World Administrative Radio Conference
WMO. World Meteorological Organization

X-SAR. synthetic aperture radar in X-band frequency range

References

1. U.S. Congress, Office of Technology Assessment, "International Cooperation and Competition in Civilian Space Activities," Report OTA-ISC-239, Washington, D.C. (July 1985).

2. National Oceanic and Atmospheric Administration and National Aeronautics and Space Administration, "Space-Based Remote Sensing of the Earth and its Atmosphere: A Report to the Congress," Washington, D.C. (September 1987).

3. Müller, E., and T. Christiansen, "Satellitendaten in Entwicklungsprojekten aus der Sicht eines im Agrarbereich tätigen Consultingunternehmens [Satellite Data in Development Projects from the Perspective of an Agricultural Consulting Company]." In: *Die Nutzung von Fernerkundungsdaten in der Bundesrepublik Deutschland* [The Utilization of Remote Sensing Data in the Federal Republic of Germany], *Proceedings* of the Bundesministerium für Forschung und Technologie Seminar, DGLR (January 1986) p. 81.

4. Gillessen, W., "Rechnergestütze Umweltplanung mit digitalen Geländedaten [Computer-Based Environmental Planning Using Digital Land Data]." In: *Die Nutzung von Fernerkundungsdaten in der Bundesrepublik Deutschland* [The Utilization of Remote Sensing Data in the Federal Republic of Germany], *Proceedings* of the Bundesministerium für Forschung und Technologie Seminar, DGLR (January 1986) p. 421.

5. Steinborn, W., "Analysis of Utilization Potential of Remote Sensing Data," Battelle Report, Frankfurt, West Germany (1987).

6. Trevett, J. W., Hunting Technical Services, Ltd., United Kingdom, personal communication (1986).

7. U.S. Congress, Office of Technology Assessment, "Remote Sensing and the Private Sector: Issues for Discussion—A Technical Memorandum," Report OTA-TM-ISC-20, Washington, D.C. (March 1984).

8. Information provided by the Earth Observation Satellite Company, Maryland, United States, February 1989. See also *EOSAT Directory of Landsat-Related Products and Services, United States Edition, 1988,* and *International Edition, 1987.*

9. U.S. Department of Commerce, "Space Commerce: An Industry Assessment," Washington, D.C. (May 1988).

10. Konecny, G., "Kartographische Nutzung von Weltraumbildern [Cartographic Utilization of Space Images]." In: *Die Nutzung von Fernerkundungsdaten in der Bundesrepublik Deutschland* [The Utilization of Remote Sensing Data in the Federal Republic of Germany], *Proceedings* of the Bundesministerium für Forschung und Technologie Seminar, DGLR (January 1986) p. 115.

11. Langner, E., Deutsche Forschungs- und Versuchsanstalt für Luft- und Raumfahrt e.V., Cologne, West Germany, personal communication (December 1986).

12. U.S. Congress, Office of Technology Assessment, "Unispace '82: A Context for International Cooperation and Competition—A Technical Memorandum," Report OTA-TM-ISC-26, Washington, D.C. (March 1983).

13. Battelle Institut e.V., "Innovation Processes and Innovation Policy: An Introduction," Special Report to German Ministry of Economics, Frankfurt, West Germany (April 1983).

14. Battelle Institut e.V., "Analysis of Existing Aids to European Industry for Development of New Products and Innovation—EXADI," Report BF-R-66.569, Frankfurt, West Germany (May 1987).

15. Organization for Economic Cooperation and Development (OECD), OECD/STIID Databank, Paris, France (July 1988).

16. White House press release (May 16, 1983).

17. U.S. Congress, Office of Technology Assessment, "Launch Options for the Future: Buyer's Guide," Report OTA-ISA-383, Washington, D.C. (July 1988).

18. U.S. Congress, Office of Technology Assessment, "Reducing Launch Operations Costs: New Technologies and Practices," Report TM-ISC-28, Washington, D.C. (August 1988).

19. White House fact sheet, Presidential Directive on National Space Policy (February 11, 1988).

20. *Aviation Week and Space Technology* (December 19, 1988) pp. 73-76.

21. U.S. Department of Commerce, *1989 U.S. Industrial Outlook, 30th Annual Edition,* Washington, D.C. (January 1989) p. 44-1.

22. *Proceedings* of the First European Workshop on Flight Opportunities for Small Payloads, Frascati, Italy, February 8-10, 1989, published by European Space Agency, Paris, France, Report ESA SP-298 (May 1989).

23. U.S. Department of Commerce, Study Contracts with KRS Remote Sensing, Egan Group, and TASC (February 1988).

24. Information from Space America, Inc. (1988).

25. Cougnet, C., "Earth Observation." In: European Space Agency, "Study on Long Term Evolution Towards European Manned Space Flight," Report Contract No. 6669/86/6/NL/PP(SC), Paris, France (October 1986).

26. Logsdon, J. M., *Space Policy* (February 1986) p. 9.

27. National Aeronautics and Space Administration, Office of Space Flight, "Payload Flight Assignments: NASA Mixed Fleet," Washington, D.C. (January 1989).

28. Conference on Remote Sensing Applications, Denver, Colorado (June 3-5, 1986).

29. United Nations General Assembly, "Report of the Committee on the Peaceful Uses of Outer Space," Official Records, 40th Session, Supplement No. 20, A/40/20, New York (1985).

30. Foders, F., *Die Weltwirtschaft,* Volume 1 (1985) p. 146.

31. Reifarth, J., Bundesministerium für Forschung und Technologie, Bonn, West Germany, personal communication (December 1986).

32. United Nations General Assembly, "International Cooperation in the Peaceful Uses of Outer Space," Official Records, 41st Session, A/SPC/41/L.29, New York, 1986.

33. Senate Committee on Aeronautical and Space Sciences, 90th Congress, 1st Session, "Treaty on Principles Governing the Activities of States in the Exploration and Use of Outer Space Including the Moon and Other Celestial Bodies," Washington, D.C. (April 1967).

34. Rivereau, J.-C., SPOT Image, Toulouse, France, personal communication (December 1986). See also *Space,* Volume 4 (1988) p. 9.

35. U.S. Congress, Office of Technology Assessment, "Civilian Space Policy and Applications," Report OTA-STI-177, Washington, D.C. (June 1982).

36. Senate Committee on Aeronautics and Space Sciences, 92nd Congress, 2nd Session, Staff Report, "Convention on International Liability for Damage Caused by Space Objects," Washington, D.C. (March 1972).

37. *Aviation Week and Space Technology* (September 12, 1988) p. 30.

38. Order (EWG) No. 2137/85 of the Council Meeting on July 7, 1985 on the Founding of a European Economic Interest Group (EWIV), Official Gazette of the European Community, No. L199 (July 1985) p. 1.

39. Battelle, "Outside Users' Payload Model," Report to NASA, Battelle Columbus Laboratories, Columbus, Ohio (July 1985). Reports are updated annually.

40. U.S. Congress, Commercial Space Launch Act of 1984, No. 2601, Washington, D.C. (1984).

41. Ministry of Research and Technology (ed.), "Documents Related to the Cabinet Decision of the German Government on Space Policy," Bundesministerium für Forschung und Technologie No. 4/85, Bonn, West Germany (1985).

42. Preuß, K.-H., and R. H. Simen (eds.), "Space Research in the Federal Republic of Germany," Inter Nationes, Bonn (1987).

43. Deutsche Forschungs- und Versuchsanstalt für Luft- und Raumfahrt e.V. (DFVLR), "Research and Development Program 1986," Cologne, West Germany (January 1986). See also DFVLR Annual Reports.

44. Information from Deutsche Forschungsanstalt für Luft- und Raumfahrt e.V. (DLR), Cologne, West Germany (December 1988).

45. Schlude, H., German Remote Sensing Data Center, Oberpfaffenhofen, West Germany, personal communication (December 1986).

46. Deutsche Forschungs- und Versuchsanstalt für Luft- und Raumfahrt e.V., Division of Project Management PT-WF, "BMFT Support Program; Acquisition and Utilization of Satellite Data for Earth Observation, for Earth and Ocean Physics, and for Atmospheric Research," Semi-Annual Management Report, Cologne, West Germany (January–June 1985).

47. European Space Directory 1989, published by Sevig Press, Paris, France (March 1989).

48. European Space Agency Council, "Optional Programmes of the Agency, Legal Data Sheets," ESA/C(88)2, Rev. 1., European Space Agency, Paris, France (July 1988).

49. Centre National d'Études Spatiales, "l'Espace en France," Business Pamphlet, Paris, France (December 1985).

50. Information from Prospace, Paris, France (December 1988).

51. Centre National d'Études Spatiales, "Annual Report 1987," Paris, France (1988).

52. Sacotte, D., Centre National d'Études Spatiales, Paris, France, personal communication (December 1986).

53. Centre National d'Études Spatiales, "Budgets and Programs of the Centre National d'Études Spatiales," Paris, France (1988).

54. Ricottilli, M., Telespazio, Rome, Italy, personal communication (December 1986).

55. Piano Spaziale Nazionale, External Relations (December 1987).

56. Goetz, J., German Embassy, Rome, Italy, personal communication (December 1986).

57. British National Space Centre, "Britain in Space" (brochure), London, United Kingdom (October 1986).

58. Lodge, D., British National Space Centre, London, United Kingdom, personal communication (December 1986).

59. Space, Volume 4 (September–October 1988) p. 44.

60. Archer, P., Nigel Press Associates, Kent, United Kingdom, personal communication (December 1986).

61. European Space Agency, "Forward to the Future," Paris, France (October 1986).

62. Space, Volume 4 (May–June 1988), p. 34.

63. European Space Agency Bulletin (February 1989), p. 101.

64. European Space Agency and Arianespace, "The Guiana Space Center," business pamphlet (February 1988).

65. European Space Agency press release, Paris, France (February 13, 1987).

66. *European Space Agency Annual Report 1987*, published by European Space Agency Publications Division, ESTEC Noordwijk, Netherlands (May 1988).

67. Dornier System GmbH, Friedrichshafen, West Germany, "ERS-1" technical brochure (1986).

68. European Space Agency, "Looking Down—Looking Forward: Earth Observation—Science and Applications, A Perspective," Special Publication ESA-SP1073, European Space Agency, Paris, France (January 1985).

69. European Space Agency Council, "Expenditure for ESA Long Term Plan, All Programmes 1987–2000," ESA/C(87)3, European Space Agency, Paris, France (1987).

70. European Space Agency Administrative and Finance Committee, Document ESA/AF(87)12, add. 6, Paris, France (June 25, 1987).

71. Information from Arianespace, Paris, France (December 1988).

72. Furniss, T., "Satellite Launcher World Market to the End of the Century." In: *Space Commerce '88, 2nd International Conference on the Commercial and Industrial Uses of Outer Space*, published by Gordon and Breach, Montreux, Switzerland (August 1988) p. 240.

73. Pyke, T. N., "Satellite Remote Sensing for Resources Development." In: *Space Commerce '88, 2nd International Conference on the Commercial and Industrial Uses of Outer Space*, published by Gordon and Breach, Montreux, Switzerland (August 1988) p. 157.

74. National Aeronautics and Space Administration, Office of Commercial Programs, "Accessing Space: A Catalogue of Processes, Equipment and Resources for Commercial Users," Special Publication NASA NP118, National Aeronautics and Space Administration, Washington, D.C. (September 1988).

75. Battelle, "Sources for Landsat Assistance and Services," Earth Resources Satellite Data Applications Series, Report to National Aeronautics and Space Administration, Module U-5, Battelle Columbus Laboratories, Columbus, Ohio (January 1980).

76. Cotter, D., and I. Wolzer (eds.), "Federal Agency Satellite Requirements, Envirosat 2000 Report," U.S. Department of Commerce and National Oceanic and Atmospheric Administration, Washington, D.C. (July 1985).

77. National Aeronautics and Space Administration, Budget Estimates, Fiscal Year 1989, Washington, D.C. (1988).

78. United States Congress, Congressional Budget Office, "The NASA Program in the 1990s and Beyond: A Special Study," Washington, D.C. (May 1988).

79. *Space Commerce Bulletin,* Volume 3 (August 1986) p. 7.

80. Space Activities Commission, "Outline of Japan's Space Development Policy," Tokyo, Japan (1978, revised 1984). See also SAC Reports "Future Space Activities in Japan" (May 1987) and "Space Development Program" (March 1988, revised August 1988).

81. National Space Development Agency of Japan, "Outline of H-II Rocket," technical brochure, Tokyo, Japan (1986).

82. National Space Development Agency of Japan, "Marine Observation Satellite-1," technical brochure, Tokyo, Japan (1987).

83. National Space Development Agency of Japan, "Earth Resources Satellite-1," technical brochure, Tokyo, Japan (1987).

84. Sahm, P. R., R. Jansen, and M. H. Keller (eds.), *Proceedings of the Symposium, Scientific Results of the German Mission D1*, Nordeney, West Germany (August 27–29, 1986). See also *Naturwissenschaften*, Volume 7 (July 1986).

85. Pardoe, G., "The Peaceful Uses of Space," *Phys. Technol.*, Volume 16 (1985) p. 100.

86. National Research Council Report, "Industrial Applications of the Microgravity Environment," Washington, D.C. (May 1988).

87. Diehl, R., "Künstliche Einkristalle als Werkstoffe," *Fortschr. Miner.*, Volume 63 (1985) p. 207.

88. Aβmus, W., Institute for Crystallography, University of Frankfurt, West Germany, personal communication (March 1986).

89. Beneking, H., Institute for Semiconductor Technology, RWTH Aachen, Aachen, West Germany, personal communication (April 1986).

90. *Aviation Week and Space Technology* (December 19, 1988) pp. 56, 95.

91. Walter, H. U. (ed.), "Fluid Sciences and Materials Science in Space: A European Perspective," Springer-Verlag, Berlin (June 1987).

92. Finke, W., "Issues and Discussion for the Industrial Utilization of Space," Bundesministerium für Forschung und Technologie paper, Bonn, West Germany (June 1985).

93. Information from INTOSPACE, Hannover, West Germany (January 1989).

94. Esterle, A., "Manufacturing in Space." In: *Space Commerce '86, International Conference on the Commercial and Industrial Uses of Outer Space*, Montreux, Switzerland (June 16-20, 1986). Published by Interavia Publishing Group, Switzerland (1986) p. 81.

95. Monti, R., University of Naples, Italy, personal communication (December 1986). See also, "Microgravity Advanced Research and Support (MARS) Center," technical brochure, Naples, Italy (1988).

96. National Aeronautics and Space Administration, Microgravity Materials Science Assessment Task Force, Final Report, Washington, D.C. (June 1987).

97. Hieronimus, A. M., European Space Agency, Paris, France, personal communication (July 1986).

98. National Space Development Agency of Japan, "FMPT First Material Processing Test," engineering information manual, Tokyo, Japan (October 1986).

99. *Science and Technology in Japan* (August 1987) p. 6.

100. Ishikawa, M., Mitsubishi Research Institute, Tokyo, Japan, personal communication (September 1987 and March 1989).

101. Battelle Institut e.V., "Study of the Utilization of the Microgravity Environment with Analysis of Failures in Micro-G Experiments," Special Report, Frankfurt, West Germany (February 1987).

102. "Space in Japan, 1986-1987" (booklet), published by Keidanren (Federation of Economic Organizations), Tokyo, Japan (November 1987).

103. Japan Space Utilization Promotion Center, In Space '88 Symposium, Tokyo, Japan (November 29-30, 1988). These symposia have been held annually in Tokyo since 1985.

104. National Aeronautics and Space Administration, Office of Commercial Programs, "A Progress Report 1988," Washington, D.C. (January 1989).

105. Space Station Commercial User Workshop, National Aeronautics and Space Administration Headquarters, Washington, D.C. (October 1988).

106. Kohli, R., "Survey of U.S. Companies Involved in Space Materials Processing Activities," Battelle Special Report, Columbus, Ohio (October 1988).

107. *Space*, Volume 4 (May-June 1988) p. 4.

108. British Aerospace, "HOTOL," business pamphlet, London (October 1988).

109. Council Meeting of the European Space Agency, The Hague, Netherlands (November 9-10, 1987).

110. Waltz, D. M., "Is There Business in Space? Outlook for Commercial Space Materials Processing," presented at the American Institute of Astronautics and Aeronautics Annual Meeting, Long Beach, California (1981).

111. U.S. Congress, Congressional Budget Office, "Using Federal R&D to Promote Commercial Innovation," Washington, D.C. (April 1988).

112. Harr, M., "Materials Processing in Space." In: European Space Agency, "Study on Long Term Evolution Towards European Manned Space Flight," Report Contract No. 6669/86/NL/PP(SC), Paris, France (October 1986).

113. European Space Agency, *European Utilization Aspects of Low Earth Orbit Space Station Elements, Phase III,* Volumes I, II, and III, Report Contract No. 5/234/82/F/FC(SC), Rider No. 2, Paris, France (1985).

114. Kienbaum Unternehmungsberatung GmbH, "Methods for Industrial Marketing of the Possibilities of Space," Düsseldorf, West Germany (March 1985).

115. Heise, O., "Perspectives on Space Utilization for Industry," presented at the Industrial Roundtable, German Industries Association (January 1986).

116. *U.S. Federal Register* (August 14, 1979).

117. *Space Business News* (December 30, 1985).

118. *Space Commerce Bulletin* (December 20, 1985).

119. The Japan Key Technology Center, business pamphlet, Tokyo (1987).

120. Kodak Limited, "Fluid Physics in Space: The Kodak Limited Experiment Aboard Spacelab-1," technical brochure, United Kingdom (1983).

121. Wolbers, H. L., "Space Station Needs, Attributes, and Architectural Options," Report MDC H0532A, McDonnell-Douglas Corporation, California (April 1983).

122. Ratafia, M., *Biotechnology,* Volume 5 (1987) p. 692.

123. National Aeronautics and Space Administration, Microgravity Science and Applications Division, "A Program Overview: 1986–87," Washington, D.C. (May 1988).

124. Bryskiewicz, T., *Journal of Crystal Growth,* Volume 85 (1987) p. 136.

125. Doblmaier, T., P. Grafing, and K. Knight, "Preliminary Market Analysis for Large Zeolite Crystals," Report CR-182882, National Aeronautics and Space Administration, Washington, D.C. (October 1987).

126. National Commission on Space, *Pioneering the Space Frontier,* Bantam Books, New York (1986). National Aeronautics and Space Administration, "Leadership and America's Future in Space", Washington, D.C. (August 1987).

127. Ersfeld, H., "Legal Problems in the Construction of Space Stations," presented at the International Colloquium on Economic Utilization of Space Stations, Hannover, West Germany (June 1986).

128. U.S. Congress, Office of Technology Assessment, "Space Stations and the Law: Selected Legal Issues—Background Paper," Report OTA-BP-ISC-41, Washington, D.C. (August 1986).

129. Dhabi, M. B., "The Space Insurance Market: Current Situation and Prospects." In: *Space Commerce '86, International Conference on the Commercial and Industrial Uses of Outer Space,* Montreux, Switzerland (June 16–20, 1986). Published by Interavia Publishing Group, Switzerland (1986) p. 141.

130. Dhabi, M. B., "Considerations on Satellite Liability Insurance." In: *Space Commerce '88, 2nd International Conference on the Commercial and Industrial Uses of Outer Space,* published by Gordon and Breach, Montreux, Switzerland (August 1988), p. 421.

131. National Aeronautics and Space Administration, "Commercial Space Policy and Implementation Plan," Washington, D.C. (October 1984).

132. Koelle, D. E., H. Kuczera, and E. Högenauer, "The German SÄNGER Space Transportation System Concept." In: *Proceedings of the 16th International Symposium on Space Technology and Science* (May 22–27, 1988) Sapporo, Japan.

133. Steinborn, W., "The German Program on Materials Science in Space." In: *Barth Orient Applications of Space Technology,* Volume 1 (1986) p. 113.

134. Kohli, R., "Analysis of Existing Aids to U.S. Industry for New Products and Innovations," Battelle Special Report, Columbus, Ohio (December 1986).

135. Avduyevsky, V. S. (ed.), "Manufacturing in Space: Processing Problems and Advances." In: *Advances in Science and Technology in the USSR,* MIR Publishers, Moscow (1985).

136. Kohli, R., "Description of U.S. Microgravity Database Systems and Modeling Efforts," Battelle Special Report, Columbus, Ohio (February 1987).

137. National Aeronautics and Space Administration, "Spinoff 1987," Washington, D.C. (1987).

138. Deutsche Forschungs- und Versuchsanstalt für Luft- und Raumfahrt e.V., "Utilization of Weightlessness," 2nd Project Meeting, Cologne, West Germany (March 23, 1987).

139. Kayser-Threde GmbH, "Get Away Payload Structure (GAPS)," technical brochure, Munich, West Germany (1988).

140. National Aeronautics and Space Administration, "Scientist's Guide for Application to the Microgravity Materials Science Laboratory," Cleveland, Ohio, U.S.A. (September 1985).

141. National Aeronautics and Space Administration, "Centers for the Commercial Development of Space," Washington, D.C. (September 1987).

142. *Aviation Week and Space Technology* (December 19, 1988) p. 69.

143. Information from National Aeronautics and Space Administration, Office of Commercial Programs, Washington, D.C. (December 1988).

144. Furniss, T., "Quest for Microgravity," *Space*, Volume 4 (March–April 1988) p. 16, and (July–August 1988) p. 19.

145. Deutsche Forschungs- und Versuchsanstalt für Luft- und Raumfahrt e.V., "Utilization of Weightlessness," 3rd Project Meeting, Cologne, West Germany (June 1, 1987).

146. Information provided by National Aeronautics and Space Administration, Office of Commercial Programs, Washington, D.C. (February 1989).

147. Steinborn, W., Battelle Büro, Bonn, West Germany, personal communication (December 1988).

148. Information from Novespace, Paris, France (1988).

149. *Commercial Space* (Spring 1986) p. 86.

Index

Note: Items in this alphabetical index are cited exactly as they occur in text, whether abbreviated or in full form. For example, entries for "Agency for International Development" can be found under that listing and under "AID."